Frontiers in Mathematics

Advisory Editorial Board

François Bouchut

Nonlinear Stability
of Finite Volume Methods
for Hyperbolic
Conservation Laws

and Well-Balanced Schemes
for Sources

Birkhäuser Verlag
Basel • Boston • Berlin

Author's address:

François Bouchut
Département de Mathématiques et Applications
CNRS & Ecole Normale Supérieure
45, rue d'Ulm
75230 Paris cedex 05
France
e-mail: Francois.Bouchut@ens.fr

2000 Mathematical Subject Classification 76M12; 65M06

A CIP catalogue record for this book is available from the
Library of Congress, Washington D.C., USA

Bibliographic information published by Die Deutsche Bibliothek
Die Deutsche Bibliothek lists this publication in the Deutsche National-
bibliografie; detailed bibliographic data is available in the Internet at
<http://dnb.ddb.de>.

ISBN 3-7643-6665-6 Birkhäuser Verlag, Basel – Boston – Berlin

© 2004 Birkhäuser Verlag, P.O. Box 133, CH-4010 Basel, Switzerland
Part of Springer Science+Business Media
Cover design: Birgit Blohmann, Zürich, Switzerland
Printed on acid-free paper produced from chlorine-free pulp. TCF ∞
ISBN 3-7643-6665-6

www.birkhauser.ch

Contents

Preface

By writing this monograph, I would like first to provide a useful gathering of some knowledge that everybody involved in the numerical simulation of hyperbolic conservation laws could have learned in journals, in conferences communications, or simply by discussing with researchers or engineers. Most of the notions discussed along the chapters are indeed either extracted from journal articles, or are natural extensions of basic ideas introduced in these articles. At the moment I write this book, it seems that the materials concerning the subject of this book, the nonlinear stability of finite volume methods for hyperbolic systems of conservation laws, have never been put together and detailed systematically in unified notation. Indeed only the scalar case is fully developed in the existing textbooks. For this reason, I shall intentionally and systematically skip the notions that are almost restricted to scalar equations, like total variation bounds, or monotonicity properties. The most well-known system is the system of gas dynamics, and the examples I consider are all of gas dynamics type.

The presentation I make does not intend to be an extensive list of all the existing methods, but rather a development centered on a very precise aim, which is the design of schemes for which one can rigorously prove nonlinear stability properties. At the same time, I would not like this work to be a too theoretical exposition, but rather a useful guide for the engineer that needs very practical advice on how to get such desired stability properties. In this respect, the nonlinear stability criteria I consider, the preservation of invariant domains and the existence of entropy inequalities, meet this requirement. The first one enables to ensure that the computed quantities remain in the physical range: nonnegative density or energy, volume fraction between 0 and 1.... The second one is twofold: it ensures the computation of admissible discontinuities, and at the same time it provides a global stability, by the property that a quantity measuring the global size of the data should not increase. This replaces in the nonlinear context the analysis by Fourier modes for linear problems.

Again in the aim of direct applicability, I consider only fully discrete explicit schemes. The main subject is therefore the study of first-order Godunov-type schemes in one dimension, and in the analysis it is always taken care of the suitable CFL condition that is necessary. I nevertheless describe a classical second-order extension method that has the nonlinear stability property we are especially interested in here, and also the usual procedure to apply the one-dimensional solvers to multi-dimensional problems interface by interface.

When establishing rigorous stability properties, the difficulty to face is not to put too much numerical diffusion, that would definitely remove any practical interest in the scheme. In this respect, in the Godunov approach, the best choice is the exact Riemann solver. However, it is computationally extremely expensive, especially for systems with large dimension. For this reason, it is necessary to design fast solvers that have minimal diffusion when the computed solution has some features that need especially be captured. This is the case when one wants to compute contact discontinuities. Indeed these discontinuities are the most diffused ones, since they do not take benefit of any spatial compression phenomena that occurs in shock waves. This is the reason why, in the first part of the monograph, I especially make emphasis on these waves, and completely disregard shock waves and rarefaction waves, the latter being indeed continuous. There has been an important progress over the last years concerning the justification of the stability of solvers that have minimal diffusion on contact discontinuities, similar as in the exact Riemann solver. I especially detail the approach by relaxation, that is extremely adapted to this aim, with the most recent developments that underly the resolution of a quasilinear approximate system with only linearly degenerate eigenvalues. This seems to be a very interesting level of simplification of a general nonlinear system, which allows better properties than the methods involving only a purely linear system, like the Roe method or the kinetic method. I indeed provide a presentation that progressively explains the different approaches, from the most general to the most particular. Kinetic schemes form a particular class in relaxation schemes, that form a particular class in approximate Riemann solvers, that lead themselves to a particular class of numerical fluxes.

The second part of the monograph is devoted to the numerical treatment of source terms that can appear additionally in hyperbolic conservation laws. This problem has been the object of intensive studies recently, at the level of analysis with the occurrence of the resonance phenomenon, as well as at the level of numerical methods. The numerical difficulty here is to treat the differential term and the source as a whole, in such a way that the well-balanced property is achieved, which is the preservation with respect to time of some particular steady states exactly at the discrete level. This topic is indeed related to the above described difficulty associated to contact discontinuities. In this second part of the book, my intention is to provide a systematic study in this context, with the extension of the notions of invariant domains, entropy inequalities, and approximate Riemann solvers. The consistency is quite subtle with sources, because a particularity of unsplit schemes is that they are not written in conservative form. This leads to a difficulty in justifying the consistency, and I explain this topic very precisely, including at second order and in multidimension. I present several methods that have been proposed in the literature, mainly for the Saint Venant problem which is the typical system with source having this difficulty of preserving steady states. They are compared concerning positivity and concerning the ability to treat resonant data. In particular, I provide a detailed analysis of the hydrostatic reconstruction method, which is extremely interesting because of its simplicity and stability properties.

I wish to thank especially F. Coquel, B. Perthame, L. Gosse, A. Vasseur, C. Simeoni, T. Katsaounis, M.-O. Bristeau, E. Audusse, N. Seguin, who enabled me to understand many things, and contributed a lot in this way to the existence of this monograph.

Paris, March 2004   François Bouchut

Chapter 1

Quasilinear systems and conservation laws

Our aim is not to develop here a full theory of the Cauchy problem for hyperbolic systems. We would like rather to introduce a few concepts that will be useful in our analysis, from a practical point of view. For more details the interested reader can consult [91], [92], [31], [44], [45], [33].

1.1 Quasilinear systems

A one-dimensional first-order *quasilinear system* is a system of partial differential equations of the form

$$\partial_t U + A(U)\partial_x U = 0, \qquad t > 0, \ x \in \mathbb{R}, \tag{1.1}$$

where $U(t,x)$ is a vector with p components, $U(t,x) \in \mathbb{R}^p$, and $A(U)$ is a $p \times p$ matrix, assumed to be smoothly dependent on U. The system is completed with an initial data

$$U(0,x) = U^0(x). \tag{1.2}$$

An important property of the system (1.1) is that its form is invariant under any smooth change of variable $V = \varphi(U)$. It becomes

$$\partial_t V + B(V)\partial_x V = 0, \tag{1.3}$$

with

$$B(V) = \varphi'(U)A(U)\varphi'(U)^{-1}. \tag{1.4}$$

The system (1.1) is said *hyperbolic* if for any U, $A(U)$ is diagonalizable, which means that it has only real eigenvalues, and a full set of eigenvectors. According to (1.4), this property is invariant under any nonlinear change of variables. We shall only consider in this presentation systems that are hyperbolic. Let us denote the distinct eigenvalues of $A(U)$ by

$$\lambda_1(U) < \cdots < \lambda_r(U). \tag{1.5}$$

The system is called *strictly hyperbolic* if all eigenvalues have simple multiplicity. We shall assume that the eigenvalues $\lambda_j(U)$ depend smoothly on U, and have constant multiplicity. In particular, this implies that the eigenvalues cannot cross.

Then, the eigenvalue $\lambda_j(U)$ is *genuinely nonlinear* if it has multiplicity one and if, denoting by $r_j(U)$ an associated eigenvector of $A(U)$, one has for all U

$$\partial_U \lambda_j(U) \cdot r_j(U) \neq 0. \tag{1.6}$$

The eigenvalue $\lambda_j(U)$ is *linearly degenerate* if for all U

$$\forall r \in \ker\left(A(U) - \lambda_j(U)\,\mathrm{Id}\right), \quad \partial_U \lambda_j(U) \cdot r = 0. \tag{1.7}$$

Again, according to (1.4), these notions are easily seen to be invariant under nonlinear change of variables.

1.2   Conservative systems

It is well known that for quasilinear systems, the solution U naturally develops discontinuities (shock waves). The main difficulty in such systems is therefore to give a sense to (1.1). Since $\partial_x U$ contains some Dirac distributions, and $A(U)$ is discontinuous in general, the product $A(U) \times \partial_x U$ can be defined in many different ways, leading to different notions of solutions. This difficulty is somehow solved when we consider *conservative systems*, also called *systems of conservation laws*, which means that they can be put in the form

$$\partial_t U + \partial_x(F(U)) = 0, \tag{1.8}$$

for some nonlinearity F that takes values in $\mathbb{R}^p$. In other words, it means that A takes the form of a jacobian matrix, $A(U) = F'(U)$. However, this property is not invariant under change of variables. Then, a *weak solution* for (1.8) is defined to be any possibly discontinuous function U satisfying (1.8) in the sense of distributions, see for example [44], [45]. The variable U in which the system takes the form (1.8) is called the *conservative variable*.

Example 1.1. The system of isentropic gas dynamics in eulerian coordinates reads as

$$\begin{cases} \partial_t \rho + \partial_x(\rho u) = 0, \\ \partial_t(\rho u) + \partial_x(\rho u^2 + p(\rho)) = 0, \end{cases} \tag{1.9}$$

where $\rho(t, x) \geq 0$ is the density, $u(t, x) \in \mathbb{R}$ is the velocity, and the pressure law $p(\rho)$ is assumed to be increasing,

$$p'(\rho) > 0. \tag{1.10}$$

One can check easily that this conservative system is hyperbolic under condition (1.10), with eigenvalues $\lambda_1 = u - \sqrt{p'(\rho)}$, $\lambda_2 = u + \sqrt{p'(\rho)}$.

Example 1.2. The system of full gas dynamics in eulerian coordinates reads

$$\begin{cases} \partial_t \rho + \partial_x(\rho u) = 0, \\ \partial_t(\rho u) + \partial_x(\rho u^2 + p) = 0, \\ \partial_t(\rho(u^2/2 + e)) + \partial_x((\rho(u^2/2 + e) + p)u) = 0, \end{cases} \tag{1.11}$$

where $\rho(t, x) \geq 0$ is the density, $u(t, x) \in \mathbb{R}$ is the velocity, $e(t, x) > 0$ is the internal energy, and $p = p(\rho, e)$. Thermodynamic considerations lead to assume that

$$de + p\,d(1/\rho) = T\,ds, \qquad (1.12)$$

for some temperature $T(\rho, e) > 0$, and specific entropy $s(\rho, e)$. Taking then (ρ, s) as variables, the hyperbolicity condition is (see [45])

$$\left(\frac{\partial p}{\partial \rho} \right)_s > 0, \qquad (1.13)$$

where the index s means that the derivative is taken at s constant. The eigenvalues are $\lambda_1 = u - \sqrt{\left(\frac{\partial p}{\partial \rho} \right)_s}$, $\lambda_2 = u$, $\lambda_3 = u + \sqrt{\left(\frac{\partial p}{\partial \rho} \right)_s}$, and $\sqrt{\left(\frac{\partial p}{\partial \rho} \right)_s}$ is called the sound speed.

An important point is that the equations (1.11) can be combined to give

$$\partial_t s + u\,\partial_x s = 0. \qquad (1.14)$$

This can be obtained by following the lines of (1.28)–(1.32). Thus smooth solutions of the isentropic system (1.9) can be viewed as special solutions of (1.11) where s is constant.

The discontinuous weak solutions of (1.8) can be characterized by the so called *Rankine–Hugoniot* jump relation.

Lemma 1.1. *Let $\mathcal{C}$ be a C^1 curve in $\mathbb{R}^2$ defined by $x = \xi(t)$, $\xi \in C^1$, that cuts the open set $\Omega \subset \mathbb{R}^2$ in two open sets Ω_- and Ω_+, defined respectively by $x < \xi(t)$ and $x > \xi(t)$ (see Figure 1.1). Consider a function U defined on Ω that is of class C^1 in $\overline{\Omega}_-$ and in $\overline{\Omega}_+$. Then U solves (1.8) in the sense of distributions in Ω if and only if U is a classical solution in Ω_- and Ω_+, and the Rankine–Hugoniot jump relation*

$$F(U_+) - F(U_-) = \dot{\xi}\,(U_+ - U_-) \qquad on\ \mathcal{C} \cap \Omega \qquad (1.15)$$

is satisfied, where $U_{\mp}$ are the values of U on each side of $\mathcal{C}$.

Proof. We can write

$$U = U_- \mathbb{1}_{x<\xi(t)} + U_+ \mathbb{1}_{x>\xi(t)}, \qquad F(U) = F(U_-)\mathbb{1}_{x<\xi(t)} + F(U_+)\mathbb{1}_{x>\xi(t)}. \qquad (1.16)$$

This gives

$$\begin{aligned}
\partial_t U &= (\partial_t U_-)\mathbb{1}_{x<\xi(t)} + (\partial_t U_+)\mathbb{1}_{x>\xi(t)} \\
&\quad + U_- \dot{\xi}(t)\delta(\xi(t) - x) - U_+ \dot{\xi}(t)\delta(x - \xi(t)), \\
\partial_x F(U) &= \partial_x F(U_-)\mathbb{1}_{x<\xi(t)} + \partial_x F(U_+)\mathbb{1}_{x>\xi(t)} \\
&\quad - F(U_-)\delta(\xi(t) - x) + F(U_+)\delta(x - \xi(t)),
\end{aligned} \qquad (1.17)$$

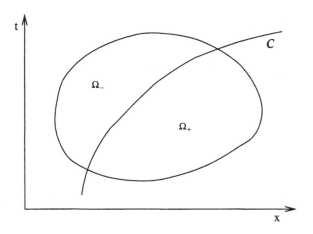

Figure 1.1: Curve $\mathcal{C}$ cutting Ω in Ω_- and Ω_+

thus

$$\partial_t U + \partial_x F(U) = (\partial_t U_- + \partial_x F(U_-))\, \mathbb{1}_{x<\xi(t)} + (\partial_t U_+ + \partial_x F(U_+))\, \mathbb{1}_{x>\xi(t)}$$
$$+ \Big(F(U_+) - F(U_-) - \dot{\xi}(t)(U_+ - U_-)\Big)\, \delta(x - \xi(t)),$$

$$(1.18)$$

and this concludes the result.   □

1.3   Invariant domains

The notion of invariant domain plays an important role in the resolution of a system of conservation laws. We say that a *convex set* $\mathcal{U} \subset \mathbb{R}^p$ is an *invariant domain* for (1.8) if it has the property that

$$U^0(x) \in \mathcal{U} \text{ for all } x \quad \Rightarrow \quad U(t,x) \in \mathcal{U} \text{ for all } x,t. \tag{1.19}$$

Notice that the convexity property is with respect to the conservative variable U. There is a full theory that enables to determine the invariant domains of a system of conservation laws. Here we are just going to assume known such invariant domain, and we refer to [92] for the theory.

Example 1.3. For a scalar law (p=1), any closed interval is an invariant domain.

Example 1.4. For the system of isentropic gas dynamics (1.9), the set $\mathcal{U} = \{U = (\rho, \rho u); \rho \geq 0\}$ is an invariant domain. It is also true that whenever $\dfrac{d(\rho\sqrt{p'(\rho)})}{d\rho} \geq 0$, the sets

$$\{(\rho, \rho u)\ ;\ u + \varphi(\rho) \leq c\}, \quad \{(\rho, \rho u)\ ;\ u - \varphi(\rho) \geq c\}, \tag{1.20}$$

are convex invariant domains for any constant c, with

$$\varphi'(\rho) = \frac{\sqrt{p'(\rho)}}{\rho}. \tag{1.21}$$

The convexity can be seen by observing that the function $(\rho, \rho u) \mapsto \rho\varphi(\rho) \pm \rho u \mp c\rho$ is convex under the above assumption.

Example 1.5. For the full gas dynamics system (1.11), the set where $e > 0$ is an invariant domain (check that this set is convex with respect to the conservative variables $(\rho, \rho u, \rho(u^2/2 + e))$).

The property for a scheme to preserve an invariant domain is an important issue of stability, as can be easily understood. In particular, the occurrence of negative values for density of for internal energy in gas dynamics calculations leads rapidly to breakdown in the computation.

1.4 Entropy

A companion notion of stability for numerical schemes is deduced from the existence of an entropy. By definition, an *entropy* for the quasilinear system (1.1) is a function $\eta(U)$ with real values such that there exists another real valued function $G(U)$, called the entropy flux, satisfying

$$G'(U) = \eta'(U)A(U), \tag{1.22}$$

where prime denotes differentiation with respect to U. In other words, $\eta'A$ needs to be an exact differential form. The existence of a strictly convex entropy is connected to hyperbolicity, by the following property.

Lemma 1.2. *If the conservative system (1.8) has a strictly convex entropy, then it is hyperbolic.*

Proof. Since η is an entropy, $\eta'F'$ is an exact differential form, which can be expressed by the fact that $(\eta'F')'$ is symmetric. Writing $(\eta'F')' = (F')^t\eta'' + \eta'F''$, the fact that F'' is itself symmetric implies that $(F')^t\eta''$ is symmetric. Since η'' is positive definite, this can be interpreted by the property that F' is self-adjoint for the scalar product defined by η''. As is well-known, any self-adjoint operator is diagonalizable, which proves the hyperbolicity. Moreover we can even conclude a more precise result: there is an orthogonal basis for η'' in which F' is diagonal. $\square$

The existence of an entropy enables, by multiplying (1.1) by $\eta'(U)$, to establish another conservation law $\partial_t(\eta(U)) + \partial_x(G(U)) = 0$. However, since we consider discontinuous functions $U(t, x)$, this identity cannot be satisfied. Instead, one should have whenever η is *convex*,

$$\partial_t(\eta(U)) + \partial_x(G(U)) \leq 0. \tag{1.23}$$

A weak solution $U(t,x)$ of (1.8) is said to be *entropy satisfying* if (1.23) holds. This property is indeed a criteria to select a unique solution to the system, that can have many weak solutions otherwise. Other criteria can be used also, but they are practically difficult to consider in numerical methods, see [45]. In the case of a piecewise C^1 function U, as in Lemma 1.1, the entropy inequality (1.23) is characterized by the Rankine–Hugoniot inequality

$$G(U_+) - G(U_-) \leq \dot{\xi}\left(\eta(U_+) - \eta(U_-)\right) \qquad \text{on } \mathcal{C} \cap \Omega. \qquad (1.24)$$

A practical method to prove that a function η is an entropy is to try to establish a conservative identity $\partial_t(\eta(U)) + \partial_x(G(U)) = 0$ for some function $G(U)$, for smooth solutions of (1.1). Then (1.22) follows automatically.

Example 1.6. For the isentropic gas dynamics system (1.9), a convex entropy is the physical energy, given by

$$\eta = \rho u^2/2 + \rho e(\rho), \qquad (1.25)$$

where the internal energy is defined by

$$e'(\rho) = \frac{p(\rho)}{\rho^2}. \qquad (1.26)$$

Its associated entropy flux is

$$G = \left(\rho u^2/2 + \rho e(\rho) + p(\rho)\right) u. \qquad (1.27)$$

The justification of this result is as follows. We first subtract u times the first equation in (1.9) to the second, and divide the result by ρ. It gives

$$\partial_t u + u \partial_x u + \frac{1}{\rho} \partial_x p(\rho) = 0. \qquad (1.28)$$

Multiplying then this equation by u gives

$$\partial_t(u^2/2) + u \partial_x(u^2/2) + \frac{u}{\rho} \partial_x p(\rho) = 0. \qquad (1.29)$$

Next, developing the density equation in (1.9) and multiplying by $p(\rho)/\rho^2$ gives

$$\partial_t e(\rho) + u \partial_x e(\rho) + \frac{p(\rho)}{\rho} \partial_x u = 0, \qquad (1.30)$$

so that by addition to (1.29) we get

$$\partial_t(u^2/2 + e(\rho)) + u \partial_x(u^2/2 + e(\rho)) + \frac{1}{\rho} \partial_x(p(\rho)u) = 0. \qquad (1.31)$$

Finally, multiplying this by ρ and adding to $u^2/2 + e(\rho)$ times the density equation gives

$$\partial_t(\rho(u^2/2 + e(\rho))) + \partial_x(\rho(u^2/2 + e(\rho))u + p(\rho)u) = 0, \qquad (1.32)$$

which is coherent with the formulas (1.25), (1.27). The convexity of η with respect to $(\rho, \rho u)$ is left to the reader.

Example 1.7. For the full gas dynamics system (1.11), according to (1.14) we have a family of entropies

$$\eta = \rho\,\phi(s), \qquad (1.33)$$

with entropy fluxes

$$G = \rho\,\phi(s)u, \qquad (1.34)$$

where ϕ is an arbitrary function such that η is convex with respect to the conservative variables $(\rho, \rho u, \rho(u^2/2 + e))$. One can deduce that the sets where $s \geq k$, k constant, are convex invariant domains. This is obtained by taking $\phi(s) = (k - s)_+$ (this choice has to be somehow adapted if $\eta = \rho\,\phi(s)$ is not convex). Then $\{s \geq k\} = \{\eta \leq 0\}$ is convex, and integrating (1.23) in x gives $d/dt(\int \eta\,dx) \leq 0$, telling that η has to vanish identically if it does initially.

Lemma 1.3. *A necessary condition for η in (1.33) to be convex with respect to $(\rho, \rho u, \rho(u^2/2 + e))$ is that $\phi' \leq 0$. Conversely, if $-s$ is a convex function of $(1/\rho, e)$ and if $\phi' \leq 0$ and $\phi'' \geq 0$, then η is convex.*

Proof. Applying Lemma 1.4 below, we have to check whether $\phi(s)$ is convex with respect to $(1/\rho, u, u^2/2 + e)$. Call $\tau = 1/\rho$, $E = u^2/2 + e$. We have according to (1.12) $ds = (pd\tau + de)/T = (pd\tau - udu + dE)/T$, thus

$$d\left[\phi(s)\right] = \phi'(s)ds = \frac{\phi'(s)}{T}\left(pd\tau - udu + dE\right), \qquad (1.35)$$

and the hessian of $\phi(s)$ with respect to (τ, u, E) is

$$\begin{aligned}
D^2_{\tau,u,E}\left[\phi(s)\right] &= \phi''(s)ds \otimes ds + \phi'(s)D^2_{\tau,u,E}s \\
&= \frac{\phi''(s)}{T^2}\left(pd\tau - udu + dE\right)^{\otimes 2} \\
&\quad + \phi'(s)(pd\tau - udu + dE) \otimes d\frac{1}{T} + \frac{\phi'(s)}{T}\left(d\tau \otimes dp - du \otimes du\right).
\end{aligned} \qquad (1.36)$$

Taking the value of this bilinear form at twice the vector $(0, 1, u)$ gives

$$D^2_{\tau,u,E}\left[\phi(s)\right] \cdot (0,1,u) \cdot (0,1,u) = -\frac{\phi'(s)}{T}, \qquad (1.37)$$

so that its nonnegativity implies that $\phi'(s) \leq 0$.
Conversely, from $ds = (pd\tau + de)/T$ we write that

$$D^2_{\tau,e}s = (pd\tau + de) \otimes d\frac{1}{T} + \frac{1}{T}d\tau \otimes dp, \qquad (1.38)$$

and inserting this into (1.36) gives

$$D^2_{\tau,u,E}[\phi(s)] = \phi''(s)ds \otimes ds + \phi'(s)(D^2_{\tau,e}s - du \otimes du/T), \qquad (1.39)$$

thus the result follows.   □

Lemma 1.4. *A scalar function $\eta(\rho, q)$, where $\rho > 0$ and q is a vector, is convex with respect to (ρ, q) if and only if η/ρ is convex with respect to $(1/\rho, q/\rho)$.*

Proof. Define $\tau = 1/\rho$ and $v = q/\rho$. Then we have

$$(\rho, q) = \varphi(\tau, v), \qquad (1.40)$$

with

$$\varphi(\tau, v) = (1/\tau, v/\tau). \qquad (1.41)$$

Define also $\eta/\rho = S(\tau, v)$, or equivalently

$$S(\tau, v) = \tau\eta(\varphi(\tau, v)). \qquad (1.42)$$

Then,

$$dS(\tau, v) = \eta(\varphi(\tau, v))d\tau + \tau\eta'(\varphi(\tau, v))d\varphi(\tau, v), \qquad (1.43)$$

and

$$\begin{aligned}
D^2_{\tau,v}S(\tau, v) = d\tau \otimes \Big(\eta'(\varphi(\tau, v))d\varphi(\tau, v)\Big) + \Big(\eta'(\varphi(\tau, v))d\varphi(\tau, v)\Big) \otimes d\tau \\
+ \tau\eta'(\varphi(\tau, v))D^2_{\tau,v}\varphi(\tau, v) \\
+ \tau\eta''(\varphi(\tau, v)) \cdot d\varphi(\tau, v) \cdot d\varphi(\tau, v).
\end{aligned} \qquad (1.44)$$

We compute from (1.41)

$$d\varphi(\tau, v) = (-d\tau/\tau^2, dv/\tau - vd\tau/\tau^2), \qquad (1.45)$$

$$D^2_{\tau,v}\varphi(\tau, v) = \big(2d\tau \otimes d\tau/\tau^3, -dv \otimes d\tau/\tau^2 - d\tau \otimes dv/\tau^2 + 2vd\tau \otimes d\tau/\tau^3\big). \qquad (1.46)$$

Now, denote

$$\eta'(\varphi(\tau, v)) = (\alpha, \beta). \qquad (1.47)$$

We have with (1.45)–(1.46)

$$\begin{aligned}
&d\tau \otimes \Big(\eta'(\varphi(\tau, v))d\varphi(\tau, v)\Big) + \Big(\eta'(\varphi(\tau, v))d\varphi(\tau, v)\Big) \otimes d\tau \\
&\quad + \tau\eta'(\varphi(\tau, v))D^2_{\tau,v}\varphi(\tau, v) \\
&= d\tau \otimes \left(-\frac{\alpha}{\tau^2}d\tau + \beta\frac{dv}{\tau} - \beta v\frac{d\tau}{\tau^2}\right) + \left(-\frac{\alpha}{\tau^2}d\tau + \beta\frac{dv}{\tau} - \beta v\frac{d\tau}{\tau^2}\right) \otimes d\tau \\
&\quad + \tau\left(\alpha\frac{2d\tau \otimes d\tau}{\tau^3} - \beta\frac{dv \otimes d\tau}{\tau^2} - \beta\frac{d\tau \otimes dv}{\tau^2} + 2\beta v\frac{d\tau \otimes d\tau}{\tau^3}\right) \\
&= 0,
\end{aligned} \qquad (1.48)$$

thus (1.44) gives

$$D^2_{\tau,v} S(\tau, v) = \tau \eta''(\varphi(\tau, v)) \cdot d\varphi(\tau, v) \cdot d\varphi(\tau, v). \tag{1.49}$$

Since $\tau > 0$ and $d\varphi(\tau, v)$ is invertible, we deduce that $D^2_{\tau,v} S(\tau, v)$ is nonnegative if and only if $\eta''(\varphi(\tau, v))$ is nonnegative, which gives the result. □

1.5 Riemann invariants, contact discontinuities

In this section we consider a general hyperbolic quasilinear system as defined in Section 1.1, and we wish to introduce some notions that are invariant under change of variables.

Consider an eigenvalue $\lambda_j(U)$. We say that a scalar function $w(U)$ is a (weak) *j-Riemann invariant* if for all U

$$\forall r \in \ker(A(U) - \lambda_j(U)\,\mathrm{Id}), \quad \partial_U w(U) \cdot r = 0. \tag{1.50}$$

This notion is obviously invariant under change of variables. A nonlinear function of several j-Riemann invariants is again a j-Riemann invariant. Applying the Frobenius theorem, we have the following.

Lemma 1.5. *Assume that λ_j has multiplicity 1. Then in the neighborhood of any point U_0, there exist $p - 1$ j-Riemann invariants with linearly independent differentials. Moreover, all j-Riemann invariants are then nonlinear functions of these ones.*

In the case of multiplicity $m_j > 1$ one could expect the same result with $p - m_j$ independent Riemann invariants. However this is wrong in general, because the Frobenius theorem requires some integrability conditions on the space $\ker(A(U) - \lambda_j(U)\,\mathrm{Id})$. Nevertheless, these integrability conditions are satisfied for most of the physically relevant quasilinear systems.

Consider still an eigenvalue $\lambda_j(U)$. We say that a scalar function $w(U)$ is a *strong j-Riemann invariant* if for all U $\partial_U w(U)$ is an eigenform associated to $\lambda_j(U)$, i.e.

$$\partial_U w(U)\,A = \lambda_j(U)\,\partial_U w(U). \tag{1.51}$$

Again this notion is invariant under change of variables, and any nonlinear function of several strong j-Riemann invariants is a strong j-Riemann invariant. The interest of this notion lies in the fact that it can be characterized by the property that a smooth solution $U(t, x)$ to (1.1) satisfies $\partial_t w(U) + \lambda_j(U)\partial_x w(U) = 0$. However, a system may have no strong Riemann invariant at all.

Lemma 1.6. *A function $w(U)$ is a strong j-Riemann invariant if and only if for any $k \neq j$, $w(U)$ is a weak k-Riemann invariant.*

Proof. This follows from the property that if (b_i) is a basis of eigenvectors of a diagonalizable matrix A, then its dual basis, i.e. the forms (l_r) such that $l_r b_i = \delta_{ir}$, is a basis of eigenforms of A. This is because $l_r A b_i = l_r \lambda_i b_i = \lambda_i \delta_{ir} = \lambda_r \delta_{ir}$, which gives $l_r A = \lambda_r l_r$.   □

Consider now λ_j a linearly degenerate eigenvalue. We say that two constant states U_l, U_r can be joined by a *j-contact discontinuity* if there exist some C^1 path $U(\tau)$ for τ in some interval $[\tau_1, \tau_2]$, such that

$$\begin{cases} \dfrac{dU}{d\tau}(\tau) \in \ker\left(A(U(\tau)) - \lambda_j(U(\tau))\,\mathrm{Id}\right) & \text{for } \tau_1 \leq \tau \leq \tau_2, \\ U(\tau_1) = U_l, \quad U(\tau_2) = U_r. \end{cases} \qquad (1.52)$$

The definition is again invariant under change of variables. We observe that if U_l, U_r can be joined by a j-contact discontinuity, we have for any j-Riemann invariant w, $(d/d\tau)[w(U(\tau))] = \partial_U w(U(\tau)) dU/d\tau = 0$, thus $w(U(\tau)) = cst = w(U_l) = w(U_r)$. This is true in particular for $w = \lambda_j$ which is a j-Riemann invariant since λ_j is assumed linearly degenerate.

If U_l, U_r can be joined by a j-contact discontinuity, we define a *j-contact discontinuity* to be a function $U(t, x)$ taking the values U_l and U_r respectively on each side of a straight line of slope $dx/dt = \lambda_j(U_l) = \lambda_j(U_r)$. Such a function will then be considered as a generalized solution to (1.1). Indeed it satisfies $\partial_t U + \lambda_j \partial_x U = 0$, and this definition is justified by the following lemma, that implies that if (1.1) has a conservative form, then $U(t, x)$ is a solution in the sense of distributions.

Lemma 1.7. *Assume that the quasilinear hyperbolic system (1.1) admits an entropy η, with entropy flux G. Then any contact discontinuity $U(t, x)$ associated to a linearly degenerate eigenvalue λ_j satisfies $\partial_t \eta(U) + \partial_x G(U) = 0$ in the sense of distributions.*

Proof. Let $w(U) = G(U) - \lambda_j(U)\eta(U)$. Then by (1.22) $\partial_U w = \partial_U \eta\,(A - \lambda_j\,\mathrm{Id}) - \eta\,\partial_U \lambda_j$, thus w is a j-Riemann invariant. It implies that $w(U_l) = w(U_r)$, i.e. $G(U_r) - G(U_l) = \lambda_j(\eta(U_r) - \eta(U_l))$, the desired Rankine–Hugoniot relation.   □

The j-contact discontinuities can indeed be characterized by the property that the j-Riemann invariants do not jump.

Lemma 1.8. *Let λ_j be a linearly degenerate eigenvalue of multiplicity m_j, and assume that in the neighborhood of some state U_0, there exist $p - m_j$ j-Riemann invariants with linearly independent differentials. Then two states U_l, U_r sufficiently close to U_0 can be joined by a j-contact discontinuity if and only if for any of these j-Riemann invariants, one has $w(U_l) = w(U_r)$.*

Proof. Since we have $p - m_j$ linearly independent forms $\partial_U w_n$ in the orthogonal of $\ker\left(A(U) - \lambda_j(U)\,\mathrm{Id}\right)$, they form a basis of this space. In particular, a vector r belongs to $\ker\left(A(U) - \lambda_j(U)\,\mathrm{Id}\right)$ if and only if $\partial_U w_n \cdot r = 0$ for $n = 1, \ldots, p - m_j$.

Therefore, the conditions (1.52) can be written $(d/d\tau)[w_n(U(\tau))] = 0$ for $n = 1, \ldots, p - m_j$ and $U(\tau_1) = U_l$, $U(\tau_2) = U_r$. We deduce that U_l, U_r can be joined by a j-contact discontinuity if and only if there exists some C^1 path joining U_l to U_r remaining in the set where $w_n(U) = w_n(U_l)$ for $n = 1, \ldots, p - m_j$. But since the differentials of w_n are independent, this set is a manifold of dimension m_j, thus it is locally connected, which gives the result. $\square$

Example 1.8. For the full gas dynamics system (1.11), one can check that the eigenvalue $\lambda_2 = u$ is linearly degenerate. By (1.14), s is a strong 2-Riemann invariant. Two independent weak 2-Riemann invariants are u and p.

Example 1.9. Consider a quasilinear system that can be put in the diagonal form

$$\partial_t w_j + \lambda_j \partial_x w_j = 0, \tag{1.53}$$

for some independent variables w_j, $j = 1, \ldots, r$, that can eventually be vector valued $w_j \in \mathbb{R}^{m_j}$, and some scalars $\lambda_j(w_1, \ldots, w_r)$ with $\lambda_1 < \cdots < \lambda_r$. Then in the variables $(w_1, \ldots, w_r)$, the matrix of the system is diagonal with eigenvalues λ_j of multicity m_j. Thus the system is hyperbolic, and the components of w_j are strong j-Riemann invariants. For any j we have $p - m_j$ independent weak j-Riemann invariants, that are the components of the w_k for $k \neq j$. Moreover, the eigenvalue λ_j is linearly degenerate if and only if it does not depend on w_j, $\lambda_j = \lambda_j(w_1, \ldots, w_{j-1}, w_{j+1}, \ldots, w_r)$. If this is the case, two states can be joined by a j-contact discontinuity if and only if the w_k for all $k \neq j$ do not jump.

Chapter 2

Conservative schemes

The notions introduced here can be found in [33], [44], [45], [97], [77].

Let us consider a system of conservation laws (1.8). We would like to approximate its solution $U(t, x)$, $x \in \mathbb{R}$, $t \geq 0$, by discrete values U_i^n, $i \in \mathbb{Z}$, $n \in \mathbb{N}$. In order to do so we consider a grid of points $x_{i+1/2}$, $i \in \mathbb{Z}$,

$$\cdots < x_{-1/2} < x_{1/2} < x_{3/2} < \ldots, \tag{2.1}$$

and we define the cells (or finite volumes) and their lengths

$$C_i =]x_{i-1/2}, x_{i+1/2}[, \qquad \Delta x_i = x_{i+1/2} - x_{i-1/2} > 0. \tag{2.2}$$

We shall denote also $x_i = (x_{i-1/2} + x_{i+1/2})/2$ the centers of the cells. We consider a constant timestep $\Delta t > 0$ and define the discrete times by

$$t_n = n\Delta t, \quad n \in \mathbb{N}. \tag{2.3}$$

The discrete values U_i^n intend to be approximations of the averages of the exact solutions over the cells,

$$U_i^n \simeq \frac{1}{\Delta x_i} \int_{C_i} U(t_n, x) \, dx. \tag{2.4}$$

A finite volume conservative scheme for solving (1.8) is a formula of the form

$$U_i^{n+1} - U_i^n + \frac{\Delta t}{\Delta x_i}(F_{i+1/2} - F_{i-1/2}) = 0, \tag{2.5}$$

telling how to compute the values U_i^{n+1} at the next time level, knowing the values U_i^n at time t_n. We consider here only first-order explicit three points schemes where

$$F_{i+1/2} = F(U_i^n, U_{i+1}^n). \tag{2.6}$$

The function $F(U_l, U_r) \in \mathbb{R}^p$ is called the *numerical flux*, and determines the scheme.

It is important to say that it is always necessary to impose what is called a *CFL condition* (for Courant, Friedrichs, Levy) on the timestep to prevent the blow up of the numerical values, under the form

$$\Delta t \, a \leq \Delta x_i, \quad i \in \mathbb{Z}, \tag{2.7}$$

where a is an approximation of the speed of propagation.

We shall often denote U_i instead of U_i^n, whenever there is no ambiguity.

2.1 Consistency

Many methods exist to determine a numerical flux. The two main criteria that enter in its choice are its stability properties, and the precision qualities it has, which can be measured by the amount of viscosity it produces and by the property of exact computation of particular solutions.

The consistency is the minimal property required for a scheme to ensure that we approximate the desired equation. For a conservative scheme, we define it as follows.

Definition 2.1. *We say that the scheme* (2.5)–(2.6) *is consistent with* (1.8) *if the numerical flux satisfies*

$$F(U, U) = F(U) \text{ for all } U. \tag{2.8}$$

We can see that this condition guarantees obviously that if for all i, $U_i^n = U$ a constant, then also $U_i^{n+1} = U$. A deeper motivation for this definition is the following.

Proposition 2.2. *Assume that for all i,*

$$U_i^n = \frac{1}{\Delta x_i} \int_{C_i} U(t_n, x) \, dx, \tag{2.9}$$

for some smooth solution $U(t, x)$ to (1.8)*, and define U_i^{n+1} by* (2.5)–(2.6)*. If the scheme is consistent, then for all i,*

$$U_i^{n+1} = \frac{1}{\Delta x_i} \int_{C_i} U(t_{n+1}, x) \, dx + \Delta t \left(\frac{1}{\Delta x_i} (\mathcal{F}_{i+1/2} - \mathcal{F}_{i-1/2}) \right), \tag{2.10}$$

where

$$\mathcal{F}_{i+1/2} \to 0, \tag{2.11}$$

as Δt and $\sup_i \Delta x_i$ tend to 0.

Proof. Let us integrate the equation (1.8) satisfied by $U(t, x)$ with respect to t and x over $]t_n, t_{n+1}[\times C_i$, and divide the result by Δx_i. We obtain

$$\frac{1}{\Delta x_i} \int_{C_i} U(t_{n+1}, x) \, dx - \frac{1}{\Delta x_i} \int_{C_i} U(t_n, x) \, dx + \frac{\Delta t}{\Delta x_i} (\underline{F}_{i+1/2} - \underline{F}_{i-1/2}) = 0, \tag{2.12}$$

where $\underline{F}_{i+1/2}$ is the exact flux

$$\underline{F}_{i+1/2} = \frac{1}{\Delta t} \int_{t_n}^{t_{n+1}} F\Big(U(t, x_{i+1/2}) \Big) \, dt. \tag{2.13}$$

Therefore, by subtracting (2.12) to (2.5), we get (2.10) with

$$\mathcal{F}_{i+1/2} = \underline{F}_{i+1/2} - F_{i+1/2}. \tag{2.14}$$

In order to conclude, we just observe that if the numerical flux is consistent (and Lipschitz continuous), $F_{i+1/2} = F(U_i^n, U_{i+1}^n) = F(U(t_n, x_{i+1/2})) + O(\Delta x_i) + O(\Delta x_{i+1})$, and since from (2.13) $\underline{F}_{i+1/2} = F(U(t_n, x_{i+1/2})) + O(\Delta t)$, we get $\mathcal{F}_{i+1/2} = O(\Delta t) + O(\Delta x_i) + O(\Delta x_{i+1})$. We can notice here that (2.11) holds indeed for a continuous numerical flux. □

The formulation (2.10)–(2.11) tells that we have an error of the form $(\mathcal{F}_{i+1/2} - \mathcal{F}_{i-1/2})/\Delta x_i$, which is the discrete derivative of a small term $\mathcal{F}$. It implies by discrete integration by parts that the error is small *in the weak sense*, the convergence holds only against a test function: if $U_h(t, x)$ is taken to be piecewise constant in space-time with values U_i^n, then one has as Δt and h tend to 0

$$\iint U_h(t, x)\varphi(t, x)\, dt\, dx \rightarrow \iint U(t, x)\, \varphi(t, x)\, dt\, dx, \qquad (2.15)$$

for any test function $\varphi(t, x)$ smooth with compact support. For the justification of such a property, we refer to [33].

2.2 Stability

The stability of the scheme can be analyzed in different ways, but we shall retain here the conservation of an invariant domain and the existence of a discrete entropy inequality. They are analyzed in a very similar way.

2.2.1 Invariant domains

Definition 2.3. *We say that the scheme* (2.5)–(2.6) *preserves a convex invariant domain* $\mathcal{U}$ *for* (1.8), *if under some CFL condition,*

$$U_i^n \in \mathcal{U} \text{ for all } i \quad \Rightarrow \quad U_i^{n+1} \in \mathcal{U} \text{ for all } i. \qquad (2.16)$$

A difficulty that occurs when trying to obtain (2.16) is that the three values U_{i-1}, U_i, U_{i+1} are involved in the computation of U_i^{n+1}. Interface conditions with only U_i, U_{i+1} can be written instead as follows, at the price of diminishing the CFL condition.

Definition 2.4. *We say that the numerical flux* $F(U_l, U_r)$ *preserves a convex invariant domain* $\mathcal{U}$ *for* (1.8) *by interface if for some* $\sigma_l(U_l, U_r) < 0 < \sigma_r(U_l, U_r)$,

$$U_l, U_r \in \mathcal{U} \Rightarrow \begin{cases} U_l + \dfrac{F(U_l, U_r) - F(U_l)}{\sigma_l} \in \mathcal{U}, \\ U_r + \dfrac{F(U_l, U_r) - F(U_r)}{\sigma_r} \in \mathcal{U}. \end{cases} \qquad (2.17)$$

Notice that if (2.17) holds for some σ_l, σ_r, then it also holds for $\sigma_l{}' \leq \sigma_l$ and $\sigma_r{}' \geq \sigma_r$, because of the convexity of $\mathcal{U}$ and of the formulas

$$U_l + \frac{F(U_l, U_r) - F(U_l)}{\sigma_l{}'} = \left(1 - \frac{\sigma_l}{\sigma_l{}'}\right)U_l + \frac{\sigma_l}{\sigma_l{}'}\left(U_l + \frac{F(U_l, U_r) - F(U_l)}{\sigma_l}\right),$$

$$U_r + \frac{F(U_l, U_r) - F(U_r)}{\sigma_r{}'} = \left(1 - \frac{\sigma_r}{\sigma_r{}'}\right)U_r + \frac{\sigma_r}{\sigma_r{}'}\left(U_r + \frac{F(U_l, U_r) - F(U_r)}{\sigma_r}\right).$$

$$\tag{2.18}$$

Proposition 2.5. (i) *If the scheme preserves an invariant domain $\mathcal{U}$ (Definition 2.3), then its numerical flux preserves $\mathcal{U}$ by interface (Definition 2.4), with $\sigma_l = -\Delta x_i/\Delta t$, $\sigma_r = \Delta x_{i+1}/\Delta t$.*
(ii) *If the numerical flux preserves an invariant domain $\mathcal{U}$ by interface (Definition 2.4), then the scheme preserves $\mathcal{U}$ (Definition 2.3), under the half CFL condition $|\sigma_l(U_i, U_{i+1})|\Delta t \leq \Delta x_i/2$, $\sigma_r(U_{i-1}, U_i)\Delta t \leq \Delta x_i/2$.*

Proof. For (i), apply (2.16) with $U_{i-1} = U_i = U_l$, $U_{i+1} = U_r$. We get the first line of (2.17) with $\sigma_l = -\Delta x_i/\Delta t$. Similarly, applying the inequality (2.16) corresponding to cell $i+1$ with $U_i = U_l$, $U_{i+1} = U_{i+2} = U_r$ gives the second line of (2.17) with $\sigma_r = \Delta x_{i+1}/\Delta t$. Conversely, for (ii), define the half-cell averages

$$U_{i+1/4}^{n+1-} = U_i - 2\frac{\Delta t}{\Delta x_i}(F(U_i, U_{i+1}) - F(U_i)),$$

$$U_{i-1/4}^{n+1-} = U_i - 2\frac{\Delta t}{\Delta x_i}(F(U_i) - F(U_{i-1}, U_i)).$$

$$\tag{2.19}$$

Then we have

$$U_i^{n+1} = \frac{1}{2}(U_{i-1/4}^{n+1-} + U_{i+1/4}^{n+1-}).\tag{2.20}$$

According to the remark above and since we have $\sigma_l(U_i, U_{i+1}) \geq -\Delta x_i/(2\Delta t)$ and $\sigma_r(U_{i-1}, U_i) \leq \Delta x_i/(2\Delta t)$, we can apply (2.17) successively with $U_l = U_i$, $U_r = U_{i+1}$, σ_l replaced by $-\Delta x_i/(2\Delta t)$, and with $U_l = U_{i-1}$, $U_r = U_i$, σ_r replaced by $\Delta x_i/(2\Delta t)$. This gives that $U_{i+1/4}^{n+1-}$, $U_{i-1/4}^{n+1-} \in \mathcal{U}$, thus by convexity $U_i^{n+1} \in \mathcal{U}$ also. $\square$

2.2.2   Entropy inequalities

Definition 2.6. *We say that the scheme (2.5)–(2.6) satisfies a discrete entropy inequality associated to the convex entropy η for (1.8), if there exists a numerical entropy flux function $G(U_l, U_r)$ which is consistent with the exact entropy flux (in the sense that $G(U, U) = G(U)$), such that, under some CFL condition, the discrete values computed by (2.5)–(2.6) automatically satisfy*

$$\eta(U_i^{n+1}) - \eta(U_i^n) + \frac{\Delta t}{\Delta x_i}(G_{i+1/2} - G_{i-1/2}) \leq 0,\tag{2.21}$$

with

$$G_{i+1/2} = G(U_i^n, U_{i+1}^n).\tag{2.22}$$

Definition 2.7. *We say that the numerical flux $F(U_l, U_r)$ satisfies an interface entropy inequality associated to the convex entropy η, if there exists a numerical entropy flux function $G(U_l, U_r)$ which is consistent with the exact entropy flux (in the sense that $G(U, U) = G(U)$), such that for some $\sigma_l(U_l, U_r) < 0 < \sigma_r(U_l, U_r)$,*

$$G(U_r) + \sigma_r \left[\eta \left(U_r + \frac{F(U_l, U_r) - F(U_r)}{\sigma_r} \right) - \eta(U_r) \right] \leq G(U_l, U_r), \qquad (2.23)$$

$$G(U_l, U_r) \leq G(U_l) + \sigma_l \left[\eta \left(U_l + \frac{F(U_l, U_r) - F(U_l)}{\sigma_l} \right) - \eta(U_l) \right]. \qquad (2.24)$$

Lemma 2.8. *The left-hand side of (2.23) and the right-hand side of (2.24) are nonincreasing functions of σ_r and σ_l respectively. In particular, for (2.23) and (2.24) to hold it is necessary that the inequalities obtained when $\sigma_r \to \infty$ and $\sigma_l \to -\infty$ (semi-discrete limit) hold,*

$$G(U_r) + \eta'(U_r)(F(U_l, U_r) - F(U_r)) \leq G(U_l, U_r), \qquad (2.25)$$

$$G(U_l, U_r) \leq G(U_l) + \eta'(U_l)(F(U_l, U_r) - F(U_l)). \qquad (2.26)$$

Proof. Since for any convex function S of a real variable, the ratio $(S(b) - S(a))/(b - a)$ is a nondecreasing function of a and b, we easily get the result by taking $S(a) = \eta(U_r + a(F(U_l, U_r) - F(U_r)))$ and $S(a) = \eta(U_l + a(F(U_l, U_r) - F(U_l)))$ respectively. $\qquad \Box$

Remark 2.1. In (2.23)–(2.24) (or in (2.25)–(2.26)), we only need to require that the left-hand side of the first inequality is less than the right-hand side of the second inequality, because then any value $G(U_l, U_r)$ between them will be acceptable as numerical entropy flux, since the consistency condition $G(U, U) = G(U)$ is automatically satisfied if the scheme is consistent.

Proposition 2.9. (i) *If the scheme is entropy satisfying (Definition 2.6), then its numerical flux is entropy satisfying by interface (Definition 2.7), with $\sigma_l = -\Delta x_i / \Delta t$, $\sigma_r = \Delta x_{i+1}/\Delta t$.*
(ii) *If the numerical flux is entropy satisfying by interface (Definition 2.7), then the scheme is entropy satisfying (Definition 2.6), under the half CFL condition $|\sigma_l(U_i, U_{i+1})|\Delta t \leq \Delta x_i/2$, $\sigma_r(U_{i-1}, U_i)\Delta t \leq \Delta x_i/2$.*

Proof. For (i), apply (2.21) with $U_{i-1} = U_i = U_l$, $U_{i+1} = U_r$. We get (2.24) with $\sigma_l = -\Delta x_i/\Delta t$. Similarly, applying the inequality (2.21) corresponding to cell $i+1$ with $U_i = U_l$, $U_{i+1} = U_{i+2} = U_r$ gives (2.23) with $\sigma_r = \Delta x_{i+1}/\Delta t$. Conversely, for (ii), define the half-cell averages

$$U_{i+1/4}^{n+1-} = U_i - 2\frac{\Delta t}{\Delta x_i}(F(U_i, U_{i+1}) - F(U_i)),$$

$$U_{i-1/4}^{n+1-} = U_i - 2\frac{\Delta t}{\Delta x_i}(F(U_i) - F(U_{i-1}, U_i)). \qquad (2.27)$$

Then we have

$$U_i^{n+1} = \frac{1}{2}(U_{i-1/4}^{n+1-} + U_{i+1/4}^{n+1-}), \qquad (2.28)$$

thus by convexity $\eta(U_i^{n+1}) \leq (\eta(U_{i-1/4}^{n+1-}) + \eta(U_{i+1/4}^{n+1-}))/2$. Since $\sigma_l(U_i, U_{i+1}) \geq -\Delta x_i/(2\Delta t)$ and $\sigma_r(U_{i-1}, U_i) \leq \Delta x_i/(2\Delta t)$, according to Lemma 2.8 we can apply the inequalities (2.24) with $U_l = U_i$, $U_r = U_{i+1}$, σ_l replaced by $-\Delta x_i/(2\Delta t)$, and (2.23) with $U_l = U_{i-1}$, $U_r = U_i$, σ_r replaced by $\Delta x_i/(2\Delta t)$, which give

$$G_{i+1/2} \leq G(U_i) - \frac{\Delta x_i}{2\Delta t}\left(\eta(U_{i+1/4}^{n+1-}) - \eta(U_i)\right),$$
$$\qquad (2.29)$$
$$G(U_i) + \frac{\Delta x_i}{2\Delta t}\left(\eta(U_{i-1/4}^{n+1-}) - \eta(U_i)\right) \leq G_{i-1/2}.$$

By addition this gives (2.21). $\square$

Semi-discrete entropy inequalities

Here we would like to make the link with *semi-discrete schemes*, where the time variable t is kept continuous and only the space variable x is discretized. Thus, (2.5)–(2.6) is replaced by

$$\frac{dU_i(t)}{dt} + \frac{1}{\Delta x_i}(F_{i+1/2} - F_{i-1/2}) = 0, \qquad F_{i+1/2} = F(U_i(t), U_{i+1}(t)), \qquad (2.30)$$

for some numerical flux $F(U_l, U_r)$. In this situation, a discrete entropy inequality writes

$$\frac{d}{dt}\eta(U_i(t)) + \frac{1}{\Delta x_i}(G_{i+1/2} - G_{i-1/2}) \leq 0, \qquad G_{i+1/2} = G(U_i(t), U_{i+1}(t)), \quad (2.31)$$

for some consistent numerical entropy flux $G(U_l, U_r)$, and it must hold for all solutions of (2.30) (here there is no notion of CFL condition). Multiplying (2.30) by $\eta'(U_i(t))$, it can be written equivalently

$$G_{i+1/2} - G_{i-1/2} - \eta'(U_i(t))\left(F_{i+1/2} - F_{i-1/2}\right) \leq 0. \qquad (2.32)$$

In other words, this means that for any U_{i-1}, U_i, U_{i+1},

$$G(U_i, U_{i+1}) - G(U_{i-1}, U_i) - \eta'(U_i)\left(F(U_i, U_{i+1}) - F(U_{i-1}, U_i)\right) \leq 0. \qquad (2.33)$$

Taking successively $U_{i-1} = U_l$, $U_i = U_{i+1} = U_r$, and $U_{i-1} = U_i = U_l$, $U_{i+1} = U_r$, we get (2.25)–(2.26). Conversely, if (2.25)–(2.26) hold, then taking $U_l = U_{i-1}$, $U_r = U_i$ in (2.25), and $U_l = U_i$, $U_r = U_{i+1}$ in (2.26) and combining the results we obtain (2.33). Therefore, in the semi-discrete case, the entropy condition exactly writes as (2.25)–(2.26), which means that the in-cell formulation (2.31) and the interface formulation (2.25)–(2.26) are fully equivalent, which is coherent with the limit $\Delta t \to 0$ in Proposition 2.9. As stated in Lemma 2.8, if a numerical flux satisfies a fully discrete entropy inequality, then the associated semi-discrete scheme also satisfies this property (this can be seen also directly by letting $\Delta t \to 0$ in (2.21)). However, the converse is not true. We refer to [95] for entropy inequalities for semi-discrete schemes.

2.3 Approximate Riemann solver of Harten, Lax, Van Leer

This section is devoted to an introduction to the most general tool involved in the construction of numerical schemes, the notion of approximate Riemann solver in the sense of Harten, Lax, Van Leer [56]. In fact, relaxation solvers, kinetic solvers and Roe solvers enter this framework. In the methods presented here, only the VFRoe method introduced in [24] does not.

We define the *Riemann problem* for (1.8) to be the problem of finding the solution to (1.8) with *Riemann initial data*

$$U^0(x) = \begin{cases} U_l \text{ if } x < 0, \\ U_r \text{ if } x > 0, \end{cases} \tag{2.34}$$

for two given constants U_l and U_r. By a simple scaling argument, this solution is indeed a function only of x/t.

Definition 2.10. *An approximate Riemann solver for (1.8) is a vector function $R(x/t, U_l, U_r)$ that is an approximation of the solution to the Riemann problem, in the sense that it must satisfy the consistency relation*

$$R(x/t, U, U) = U, \tag{2.35}$$

and the conservativity identity

$$F_l(U_l, U_r) = F_r(U_l, U_r), \tag{2.36}$$

where the left and right numerical fluxes are defined by

$$F_l(U_l, U_r) = F(U_l) - \int_{-\infty}^{0} \Big(R(v, U_l, U_r) - U_l \Big)\, dv,$$
$$F_r(U_l, U_r) = F(U_r) + \int_{0}^{\infty} \Big(R(v, U_l, U_r) - U_r \Big)\, dv. \tag{2.37}$$

It is called dissipative with respect to a convex entropy η for (1.8) if

$$G_r(U_l, U_r) - G_l(U_l, U_r) \le 0, \tag{2.38}$$

where

$$G_l(U_l, U_r) = G(U_l) - \int_{-\infty}^{0} \Big(\eta(R(v, U_l, U_r)) - \eta(U_l) \Big)\, dv,$$
$$G_r(U_l, U_r) = G(U_r) + \int_{0}^{\infty} \Big(\eta(R(v, U_l, U_r)) - \eta(U_r) \Big)\, dv, \tag{2.39}$$

and G is the entropy flux associated to η, $G' = \eta' F'$.

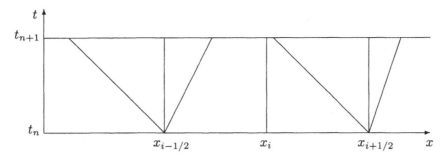

Figure 2.1: Approximate solution

It is possible to prove that the exact solution to the Riemann problem satisfies these properties. However, the above definition is rather motivated by numerical schemes. Indeed to an approximate Riemann solver we can associate a conservative numerical scheme. Let us explain how.

Consider a discrete sequence U_i^n, $i \in \mathbb{Z}$. Then we can interpret U_i^n to be the cell average of the function $U^n(x)$ which is piecewise constant over the mesh with value U_i^n in each cell C_i. In order to solve (1.8) with data $U^n(x)$ at time t_n, we can consider that close to each interface point $x_{i+1/2}$, we have to solve a translated Riemann problem. Since (1.8) is invariant under translation in time and space, we can think of sticking together the local approximate Riemann solutions $R((x - x_{i+1/2})/(t - t_n), U_i^n, U_{i+1}^n)$, at least for times such that these solutions do not interact. This is possible until time t_{n+1} under a CFL condition $1/2$, in the sense that

$$x/t < -\frac{\Delta x_i}{2\Delta t} \quad \Rightarrow \quad R(x/t, U_i, U_{i+1}) = U_i,$$

$$x/t > \frac{\Delta x_{i+1}}{2\Delta t} \quad \Rightarrow \quad R(x/t, U_i, U_{i+1}) = U_{i+1}.$$

$$(2.40)$$

Thus, as illustrated in Figure 2.1, we define an approximate solution $U(t, x)$ for $t_n \leq t < t_{n+1}$ by

$$U(t, x) = R\left(\frac{x - x_{i+1/2}}{t - t_n}, U_i^n, U_{i+1}^n\right) \quad \text{if } x_i < x < x_{i+1}. \qquad (2.41)$$

Then, we define U_i^{n+1} to be the average over C_i of this approximate solution at time $t_{n+1} - 0$. According to the definition (2.37) of F_l and F_r and by using (2.40), we get

$$U_i^{n+1} = \frac{1}{\Delta x_i} \int_{x_{i-1/2}}^{x_{i+1/2}} U(t_{n+1} - 0, x) \, dx$$

$$= \frac{1}{\Delta x_i} \int_0^{\Delta x_i/2} R(x/\Delta t, U_{i-1}^n, U_i^n) \, dx + \frac{1}{\Delta x_i} \int_{-\Delta x_i/2}^0 R(x/\Delta t, U_i^n, U_{i+1}^n) \, dx$$

$$= U_i^n + \frac{1}{\Delta x_i} \int_0^{\Delta x_i/2} \left(R(x/\Delta t, U_{i-1}^n, U_i^n) - U_i^n \right) dx$$

$$+ \frac{1}{\Delta x_i} \int_{-\Delta x_i/2}^0 \left(R(x/\Delta t, U_i^n, U_{i+1}^n) - U_i^n \right) dx$$

$$= U_i^n - \frac{\Delta t}{\Delta x_i} [F_l(U_i^n, U_{i+1}^n) - F_r(U_{i-1}^n, U_i^n)].$$

$$(2.42)$$

Therefore we see that with the conservativity assumption (2.36), this is a conservative scheme, with numerical flux

$$F(U_l, U_r) = F_l(U_l, U_r) = F_r(U_l, U_r). \tag{2.43}$$

The consistency assumption (2.35) ensures that this numerical flux is consistent, in the sense of Definition 2.1.

Remark 2.2. The approximate Riemann solver framework works as well with interface dependent solvers $R_{i+1/2}$. This is used in practice to choose a solver adapted to the data U_i, U_{i+1}, so as to produce a viscosity which is as small as possible.

Now let us examine condition (2.38). Since η is convex, we can use Jensen's inequality in (2.42), and we get

$$\eta(U_i^{n+1}) \leq \frac{1}{\Delta x_i} \int_0^{\Delta x_i/2} \eta\left(R(x/\Delta t, U_{i-1}^n, U_i^n) \right) dx$$

$$+ \frac{1}{\Delta x_i} \int_{-\Delta x_i/2}^0 \eta\left(R(x/\Delta t, U_i^n, U_{i+1}^n) \right) dx \qquad (2.44)$$

$$= \eta(U_i^n) - \frac{\Delta t}{\Delta x_i} [G_l(U_i^n, U_{i+1}^n) - G_r(U_{i-1}^n, U_i^n)].$$

Under assumption (2.38), we get

$$\eta(U_i^{n+1}) - \eta(U_i^n) + \frac{\Delta t}{\Delta x_i} [G(U_i^n, U_{i+1}^n) - G(U_{i-1}^n, U_i^n)] \leq 0, \tag{2.45}$$

for any numerical entropy flux function $G(U_l, U_r)$ such that

$$G_r(U_l, U_r) \leq G(U_l, U_r) \leq G_l(U_l, U_r), \tag{2.46}$$

thus we recover the conditions of Definition 2.6, since (2.35) ensures that this numerical entropy flux is consistent.

Another way to get (2.45) is to apply Proposition 2.9(ii). Indeed if σ_l and σ_r are chosen so that $x/t < \sigma_l \Rightarrow R(x/t, U_l, U_r) = U_l$ and $x/t > \sigma_r \Rightarrow R(x/t, U_l, U_r) = U_r$, then with (2.37) and Jensen's inequality

$$G(U_r) + \sigma_r \left[\eta \left(U_r + \frac{F_r(U_l, U_r) - F(U_r)}{\sigma_r} \right) - \eta(U_r) \right]$$
$$= G(U_r) + \sigma_r \left[\eta \left(\frac{1}{\sigma_r} \int_0^{\sigma_r} R(v, U_l, U_r)\, dv \right) - \eta(U_r) \right] \tag{2.47}$$
$$\leq G_r(U_l, U_r),$$

$$G(U_l) + \sigma_l \left[\eta \left(U_l + \frac{F_l(U_l, U_r) - F(U_l)}{\sigma_l} \right) - \eta(U_l) \right]$$
$$= G(U_l) + \sigma_l \left[\eta \left(\frac{-1}{\sigma_l} \int_{\sigma_l}^0 R(v, U_l, U_r)\, dv \right) - \eta(U_l) \right] \tag{2.48}$$
$$\geq G_l(U_l, U_r).$$

Therefore, (2.46) implies that the inequalities (2.23)-(2.24) of Definition 2.7 are satisfied, and the numerical flux is entropy satisfying by interface.

The invariant domains can also be recovered within this framework.

Proposition 2.11. *Assume that R is an approximate Riemann solver that preserves a convex invariant domain $\mathcal{U}$ for (1.8), in the sense that*

$$U_l, U_r \in \mathcal{U} \quad \Rightarrow \quad R(x/t, U_l, U_r) \in \mathcal{U} \text{ for any value of } x/t. \tag{2.49}$$

Then the numerical scheme associated to R also preserves $\mathcal{U}$ in the sense of Definition 2.3.

Proof. This is obvious with the convex formula in the first line of (2.42). Another proof is to verify that the numerical flux preserves $\mathcal{U}$ by interface, by using the convex formulas in (2.47), (2.48). □

We have seen that to any approximate Riemann solver R we can associate a conservative numerical scheme. In particular, if we use the exact Riemann solver, the scheme we get is called the *(exact) Godunov scheme*. But in practice, the exact resolution of the Riemann problem is too complicate and too expensive, especially for systems with large dimension. Thus we rather use approximate solvers. The most simple choice is the following.

2.3.1   Simple solvers

We shall call *simple solver* an approximate Riemann solver consisting of a set of finitely many simple discontinuities. This means that there exists a finite number $m \geq 1$ of speeds

$$\sigma_0 = -\infty < \sigma_1 < \cdots < \sigma_m < \sigma_{m+1} = +\infty, \tag{2.50}$$

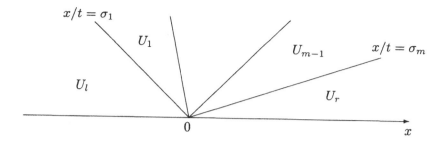

Figure 2.2: A simple solver

and intermediate states

$$U_0 = U_l, U_1, \ldots, U_{m-1}, U_m = U_r \tag{2.51}$$

(depending on U_l and U_r), such that, as illustrated in Figure 2.2,

$$R(x/t, U_l, U_r) = U_k \quad \text{if} \quad \sigma_k < x/t < \sigma_{k+1}. \tag{2.52}$$

Then the conservativity identity (2.36) becomes

$$\sum_{k=1}^{m} \sigma_k (U_k - U_{k-1}) = F(U_r) - F(U_l), \tag{2.53}$$

and the entropy inequality (2.38) becomes

$$\sum_{k=1}^{m} \sigma_k \left(\eta(U_k) - \eta(U_{k-1}) \right) \geq G(U_r) - G(U_l). \tag{2.54}$$

Conservativity thus enables to define the intermediate fluxes F_k, $k = 0, \ldots, m$, by

$$F_k - F_{k-1} = \sigma_k (U_k - U_{k-1}), \qquad F_0 = F(U_l), \ F_m = F(U_r), \tag{2.55}$$

which is a kind of generalization of the Rankine–Hugoniot relation. The numerical flux is then given by

$$F(U_l, U_r) = F_k, \quad \text{where } k \text{ is such that } \sigma_k \leq 0 \leq \sigma_{k+1}. \tag{2.56}$$

We can observe that if it happens that $\sigma_k = 0$ for some k, there is no ambiguity in this definition since (2.55) gives in this case $F_k = F_{k-1}$. An explicit formula for the numerical flux is indeed

$$\begin{aligned} F(U_l, U_r) &= F(U_l) + \sum_{\sigma_k < 0} \sigma_k (U_k - U_{k-1}) \\ &= F(U_r) - \sum_{\sigma_k > 0} \sigma_k (U_k - U_{k-1}). \end{aligned} \tag{2.57}$$

2.3.2   Roe solver

The *Roe solver* [89] is an example of simple solver. It is obtained as follows. We need first to find a $p \times p$ diagonalizable matrix $A(U_l, U_r)$ (called a *Roe matrix*), such that

$$F(U_r) - F(U_l) = A(U_l, U_r)(U_r - U_l),$$
$$A(U, U) = F'(U). \tag{2.58}$$

Then we define $R(x/t, U_l, U_r)$ to be the solution to the linear problem

$$\partial_t U + A(U_l, U_r)\partial_x U = 0, \tag{2.59}$$

with initial Riemann data (2.34). Denoting by $\sigma_1, \ldots, \sigma_m$ the distinct eigenvalues of $A(U_l, U_r)$, we can decompose $U_r - U_l$ along the eigenspaces

$$U_r - U_l = \sum_{k=1}^{m} \delta U_k, \qquad A(U_l, U_r)\delta U_k = \sigma_k \delta U_k, \tag{2.60}$$

and the solution is given by

$$R(x/t, U_l, U_r) = U_l + \sum_{k=1}^{k_0} \delta U_k, \quad \text{if } \sigma_{k_0} < x/t < \sigma_{k_0+1}. \tag{2.61}$$

This defines a simple solver, the assumption (2.58) gives indeed the conservativity (2.53), since

$$\sum_k \sigma_k \delta U_k = A(U_l, U_r)(U_r - U_l) = F(U_r) - F(U_l). \tag{2.62}$$

However, this method does generally not preserve invariant domains, and is not entropy satisfying, *entropy fixes* have to be designed. We refer the reader to the literature [45], [97], [76] for this class of schemes. For our purpose here, we shall not consider this method because it is not possible to analyze its positivity, which is a big problem when vacuum is involved.

2.3.3   CFL condition

For a simple solver we can define the local speed by

$$a(U_l, U_r) = \sup_{1 \le k \le m} |\sigma_k|. \tag{2.63}$$

Then the CFL condition (2.40) reads

$$\Delta t \, a(U_i, U_{i+1}) \le \frac{1}{2} \min(\Delta x_i, \Delta x_{i+1}). \tag{2.64}$$

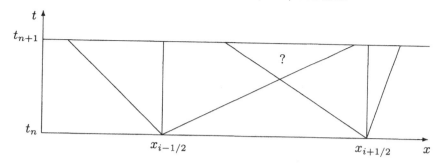

Figure 2.3: Interaction at CFL 1

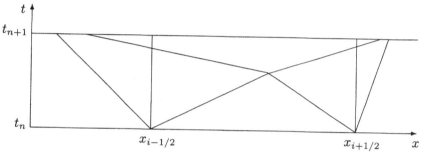

Figure 2.4: Bad interaction at CFL 1

This is called a CFL condition $1/2$. However, in practice, it is almost always possible to use a CFL 1 condition,

$$\Delta t\, a(U_i, U_{i+1}) \leq \min(\Delta x_i, \Delta x_{i+1}). \tag{2.65}$$

The reason is that since the numerical flux somehow involves only the solution on the line $x = x_{i+1/2}$ (as is seen in (2.13)), we do not really need that no interaction occurs between the Riemann problems, as was assumed in Figure 2.1. A situation like Figure 2.3 should be enough. But of course we need some kind of interaction to exist, and that the domain with question mark corresponds to acceptable values of U. A bad situation is illustrated in Figure 2.4, where even if the local problems are solved with CFL 1, the interaction produces larger speeds, and the waves attain the neighboring cells. Schemes that handle the interaction of waves at CFL larger than 1 are analyzed in [101] and the references therein.

2.3.4 Vacuum

As already mentioned, the computation of the solution to isentropic gas dynamics (1.9), or full gas dynamics (1.11) with data having vacuum is a difficult point, mainly because hyperbolicity is lost there. In the computation of an approximate

Riemann solver, if the two values U_l, U_r are vacuum data $U_l = U_r = 0$, there is no difficulty, we can simply set $R = 0$. The problem occurs when one of the two values is zero and the other is not. We shall say that an approximate Riemann solver can resolve the vacuum if in this case of two values U_l, U_r which are zero and nonzero, it gives a solution $R(x/t, U_l, U_r)$ with nonnegative density and with finite speed of propagation, otherwise the CFL condition (2.65) would give a zero timestep. The construction of solvers that are able to resolve vacuum is a main point for applications to flows in rivers with Saint Venant type equations.

2.4   Relaxation solvers

The *relaxation method* is the most recent between the ones presented here. It is used in [63], [30], [17], [11], [26]. We follow here the presentation of [27], [18] (see also [78]).

Definition 2.12. *A relaxation system for* (1.8) *is another system of conservation laws in higher dimension $q > p$,*

$$\partial_t f + \partial_x (\mathcal{A}(f)) = 0, \tag{2.66}$$

where $f(t, x) \in \mathbb{R}^q$, and $\mathcal{A}(f) \in \mathbb{R}^q$. We assume that this system is also hyperbolic. The link between (2.66) *and* (1.8) *is made by the assumption that we have a linear operator*

$$L : \mathbb{R}^q \to \mathbb{R}^p \tag{2.67}$$

and for any U, an equilibrium $M(U) \in \mathbb{R}^q$, the maxwellian equilibrium, such that for any U

$$L M(U) = U, \tag{2.68}$$

$$L \mathcal{A}(M(U)) = F(U). \tag{2.69}$$

When solving (2.66), we define

$$U \equiv Lf. \tag{2.70}$$

This cannot make any confusion since by (2.68) this gives the expected value when f is a maxwellian, $f = M(U)$.

The heart of the notion of relaxation system is the idea that $U = Lf$ should be an approximate solution to (1.8) when f solves (2.66) (exactly or approximately), and is close to maxwellian data. We have to mention that we do not consider here right-hand sides in (2.66) to achieve the relaxation to the maxwellian state, like $(M(Lf) - f)/\epsilon$, as is usual, but rather replace this by time discrete projections onto maxwellians, an approach that was introduced in [23]. It is more adapted to the numerical resolution of the conservation law (1.8) without right-hand side, see [18]. The whole process of transport in (2.66) followed by relaxation to maxwellian states, can be formalized as follows.

Proposition 2.13. *Let $\mathcal{R}(x/t, f_l, f_r)$ be an approximate Riemann solver for the relaxation system (2.66). Then*

$$R(x/t, U_l, U_r) = L \mathcal{R}\Big(x/t, M(U_l), M(U_r)\Big) \qquad (2.71)$$

is an approximate Riemann solver for (1.8).

Proof. We have obviously from the consistency of $\mathcal{R}$

$$R(x/t, U, U) = L \mathcal{R}(x/t, M(U), M(U)) = L M(U) = U, \qquad (2.72)$$

which gives the consistency of R, (2.35). Next, denote by $\mathcal{A}_l(f_l, f_r)$ and $\mathcal{A}_r(f_l, f_r)$ the left and right numerical fluxes for the relaxation system (2.66). We have

$$
\begin{aligned}
&F_l(U_l, U_r) \\
&= F(U_l) - \int_{-\infty}^{0} \Big(R(v, U_l, U_r) - U_l\Big) dv \\
&= F(U_l) - L \int_{-\infty}^{0} \Big(\mathcal{R}(v, M(U_l), M(U_r)) - M(U_l)\Big) dv \qquad (2.73) \\
&= F(U_l) + L \Big[\mathcal{A}_l(M(U_l), M(U_r)) - \mathcal{A}(M(U_l))\Big] \\
&= L \mathcal{A}_l(M(U_l), M(U_r)),
\end{aligned}
$$

and similarly

$$
\begin{aligned}
&F_r(U_l, U_r) \\
&= F(U_r) + \int_{0}^{\infty} \Big(R(v, U_l, U_r) - U_r\Big) dv \\
&= F(U_r) + L \int_{0}^{\infty} \Big(\mathcal{R}(v, M(U_l), M(U_r)) - M(U_r)\Big) dv \qquad (2.74) \\
&= F(U_r) + L \Big[\mathcal{A}_r(M(U_l), M(U_r)) - \mathcal{A}(M(U_r))\Big] \\
&= L \mathcal{A}_r(M(U_l), M(U_r)).
\end{aligned}
$$

Since $\mathcal{R}$ is conservative, $\mathcal{A}_l = \mathcal{A}_r$ and we deduce the conservativity of R (2.36), with numerical flux

$$F(U_l, U_r) = L \mathcal{A}(M(U_l), M(U_r)). \qquad (2.75)$$

Therefore the result is proved. $\qquad\square$

A very interesting property of relaxation systems is that they can handle naturally entropy inequalities, as follows. Assume that η is a convex entropy for (1.8), and denote by G its entropy flux.

Definition 2.14. *We say that the relaxation system (2.66) has an entropy extension relative to η if there exists some convex function $\mathcal{H}(f)$, which is an entropy for (2.66), which means that there exist some entropy flux $\mathcal{G}(f)$ such that*

$$\mathcal{G}' = \mathcal{H}'\mathcal{A}', \tag{2.76}$$

that these entropy and entropy-flux are extensions of the ones of the relaxed system (1.8),

$$\mathcal{H}(M(U)) = \eta(U) + cst, \tag{2.77}$$

$$\mathcal{G}(M(U)) = G(U) + cst, \tag{2.78}$$

and that the minimization principle holds,

$$\mathcal{H}(M(U)) \leq \mathcal{H}(f) \qquad whenever \ U = Lf. \tag{2.79}$$

An analysis of the occurrence of such minimization principle is provided in [18]. The interest of this notion lies in the following.

Proposition 2.15. *Assume that the relaxation system (2.66) has an entropy extension $\mathcal{H}$ relative to η, and let $\mathcal{R}(x/t, f_l, f_r)$ be an associated approximate Riemann solver, assumed to be $\mathcal{H}$ entropy satisfying. Then the approximate Riemann solver R defined by (2.71) is η entropy satisfying.*

Proof. Denote by $\mathcal{G}_l(f_l, f_r)$ and $\mathcal{G}_r(f_l, f_r)$ the left and right numerical entropy fluxes associated to $\mathcal{R}$. We have according to (2.77) and to the entropy minimization principle (2.79)

$$
\begin{aligned}
&G_l(U_l, U_r) \\
&= G(U_l) - \int_{-\infty}^{0} \Big(\eta(R(v, U_l, U_r)) - \eta(U_l) \Big) \, dv \\
&\geq G(U_l) - \int_{-\infty}^{0} \Big(\mathcal{H}(\mathcal{R}(v, M(U_l), M(U_r))) - \mathcal{H}(M(U_l)) \Big) \, dv \\
&= \mathcal{G}_l(M(U_l), M(U_r)) - \mathcal{G}(M(U_l)) + G(U_l),
\end{aligned}
\tag{2.80}
$$

and

$$
\begin{aligned}
&G_r(U_l, U_r) \\
&= G(U_r) + \int_{0}^{\infty} \Big(\eta(R(v, U_l, U_r)) - \eta(U_r) \Big) \, dv \\
&\leq G(U_r) + \int_{0}^{\infty} \Big(\mathcal{H}(\mathcal{R}(v, M(U_l), M(U_r))) - \mathcal{H}(M(U_r)) \Big) \, dv \\
&= \mathcal{G}_r(M(U_l), M(U_r)) - \mathcal{G}(M(U_r)) + G(U_r).
\end{aligned}
\tag{2.81}
$$

But because of (2.78), $-\mathcal{G}(M(U_l)) + G(U_l) = -\mathcal{G}(M(U_r)) + G(U_r)$, thus the entropy dissipativity of $\mathcal{R}$, i.e. $\mathcal{G}_r - \mathcal{G}_l \leq 0$, implies that of R, i.e. $G_r - G_l \leq 0$. $\square$

2.4.1 Nonlocal approach

The global approach to build a numerical scheme from a relaxation system is the following. We start from the piecewise constant function

$$f^n(x) = f_i^n = M(U_i^n), \quad \text{if } x_{i-1/2} < x < x_{i+1/2}, \tag{2.82}$$

and then solve

$$\partial_t f + \partial_x(\mathcal{A}(f)) = 0 \quad \text{in }]t_n, t_{n+1}[\times \mathbb{R} \tag{2.83}$$

with this initial data. Next we define

$$U(t, x) = L\, f(t, x), \quad \text{for } t_n \leq t < t_{n+1}, \tag{2.84}$$

and the new discrete values at time t_{n+1} are obtained by

$$U_i^{n+1} = \frac{1}{\Delta x_i} \int_{x_{i-1/2}}^{x_{i+1/2}} U(t_{n+1}-, x)\, dx. \tag{2.85}$$

By taking L in (2.83) and then averaging as in the proof of Proposition 2.2, we get a conservative scheme with numerical flux

$$F_{i+1/2} = \frac{1}{\Delta t} \int_{t_n}^{t_{n+1}} L\, \mathcal{A}(f(t, x_{i+1/2}))\, dt. \tag{2.86}$$

Obviously, under a CFL condition $1/2$, this is the same scheme as the one obtained from the approximate solver of Proposition 2.13, with $\mathcal{R}$ the exact solver, because $U(t, x)$ is identically the approximate solution defined in (2.41). However, the global approach has the advantage to work with CFL 1, because the wave interaction of Figure 2.3 is here exactly computed in (2.83). The only counterpart is that with this approach, we are not able to use an interface dependent solver, as stated in Remark 2.2.

Under the assumption that the relaxation system has an entropy extension $\mathcal{H}$ relative to η (Definition 2.14), we can also obtain the entropy inequality, as follows. Since f is the exact entropy solution to (2.83), we have

$$\partial_t(\mathcal{H}(f)) + \partial_x(\mathcal{G}(f)) \leq 0. \tag{2.87}$$

Integrating this inequality with respect to time and space, this gives

$$\frac{1}{\Delta x_i} \int_{x_{i-1/2}}^{x_{i+1/2}} \mathcal{H}(f(t_{n+1}-, x))\, dx - \frac{1}{\Delta x_i} \int_{x_{i-1/2}}^{x_{i+1/2}} \mathcal{H}(f(t_n, x))\, dx \\ + \frac{\Delta t}{\Delta x_i} \left(G_{i+1/2} - G_{i-1/2}\right) \leq 0, \tag{2.88}$$

with

$$G_{i+1/2} = \frac{1}{\Delta t} \int_{t_n}^{t_{n+1}} \mathcal{G}(f(t, x_{i+1/2}))\, dt, \tag{2.89}$$

which is consistent with G by (2.78). But by the minimization principle (2.79),

$$\mathcal{H}(M(U(t_{n+1}-, x))) \leq \mathcal{H}(f(t_{n+1}-, x)). \tag{2.90}$$

Finally, with (2.77), (2.82) and the Jensen inequality

$$\eta(U_i^{n+1}) \leq \frac{1}{\Delta x_i} \int_{x_{i-1/2}}^{x_{i+1/2}} \eta(U(t_{n+1}-, x)) \, dx, \tag{2.91}$$

we obtain

$$\eta(U_i^{n+1}) - \eta(U_i^n) + \frac{\Delta t}{\Delta x_i} \left(G_{i+1/2} - G_{i-1/2} \right) \leq 0. \tag{2.92}$$

2.4.2   Rusanov flux

The most simple numerical flux for solving the general system of conservation laws (1.8) is the well-known *Lax–Friedrichs* numerical flux given by

$$F(U_l, U_r) = \frac{F(U_l) + F(U_r)}{2} - c\frac{U_r - U_l}{2}, \tag{2.93}$$

where $c > 0$ is a parameter. The consistency of this numerical flux is obvious. However, the analysis of invariant domains and entropy inequalities requires a bit of work, and can be performed via a relaxation interpretation of it, that has been proposed in [63].
This relaxation system has dimension $q = 2p$, and reads

$$\begin{cases} \partial_t U + \partial_x V = 0, \\ \partial_t V + c^2 \partial_x U = 0. \end{cases} \tag{2.94}$$

Following Definition 2.12, we have here $f = (U, V)$, $\mathcal{A}(U, V) = (V, c^2 U)$, $L(U, V) = U$, $M(U) = (U, F(U))$, so that (2.68), (2.69) hold. Notice that the notation $f = (U, V)$ is coherent with the fact that we always identify U with Lf.
A slightly different way of writing (2.94) is to write it in its diagonal form,

$$\begin{cases} \partial_t(U + V/c) + c\, \partial_x(U + V/c) = 0, \\ \partial_t(U - V/c) - c\, \partial_x(U - V/c) = 0. \end{cases} \tag{2.95}$$

In this form we can rather make the (equivalent) interpretation

$$f = (f_1, f_2) = \left(\frac{U - V/c}{2}, \frac{U + V/c}{2} \right), \tag{2.96}$$

$$\mathcal{A}(f) = (-cf_1, cf_2), \qquad Lf = f_1 + f_2, \tag{2.97}$$

$$M(U) = \left(\frac{U - F(U)/c}{2}, \frac{U + F(U)/c}{2} \right), \tag{2.98}$$

for which (2.68), (2.69) is again satisfied. We can apply Proposition 2.13 with $\mathcal{R}$ the exact solver, which is given by

$$\mathcal{R}(x/t, f^l, f^r) = \begin{cases} (f_1^l, f_2^l) & \text{if } x/t < -c, \\ (f_1^r, f_2^l) & \text{if } -c < x/t < c, \\ (f_1^r, f_2^r) & \text{if } c < x/t. \end{cases} \qquad (2.99)$$

Thus (2.71) gives the simple approximate Riemann solver

$$R(x/t, U_l, U_r) = \begin{cases} U_l & \text{if } x/t < -c, \\ (U_l + U_r)/2 - (F(U_r) - F(U_l))/2c & \text{if } -c < x/t < c, \\ U_r & \text{if } c < x/t. \end{cases}$$

$$\qquad (2.100)$$

Using (2.37) or (2.75), its associated numerical flux is given by (2.93).

Now with this relaxation interpretation of the Lax–Friedrichs scheme, an analysis of entropy compatibility can be performed. A main idea is to define an extended entropy $\mathcal{H}$ with extended entropy flux $\mathcal{G}$ corresponding to an entropy η with entropy flux G of (1.8) by

$$\mathcal{H}(f) = \frac{\eta(U^-) - G(U^-)/c}{2} + \frac{\eta(U^+) + G(U^+)/c}{2},$$

$$\mathcal{G}(f) = -c\frac{\eta(U^-) - G(U^-)/c}{2} + c\frac{\eta(U^+) + G(U^+)/c}{2}, \qquad (2.101)$$

where U^-, U^+ are defined by

$$\frac{U^- - F(U^-)/c}{2} = f_1, \qquad \frac{U^+ + F(U^+)/c}{2} = f_2. \qquad (2.102)$$

This construction requires that the relations (2.102) have a solution, which means more or less that the eigenvalues $\lambda_j(U)$ of $F'(U)$ satisfy

$$|\lambda_j(U)| \leq c. \qquad (2.103)$$

This condition is called a *subcharacteristic condition*, it means that the eigenvalues of the system to be solved (1.8) lie between the eigenvalues of the relaxation system, which are $-c$ and $+c$ here. General relations between entropy conditions and subcharacteristic conditions can be found in [27] and [18]. Additional geometrical assumptions related to global convexity are indeed also necessary in order to justify the entropy inequalities. We shall not give the details here, they can be found in [17] in the more general context of flux vector splitting fluxes. Similar assumptions lead to the preservation of invariant domains, see [35], [36].

Finally, the *Rusanov flux* is obtained according to Remark 2.2 by optimizing (2.103), and taking for c in (2.93)

$$c = \sup_{U=U_l, U_r} \sup_j |\lambda_j(U)|. \qquad (2.104)$$

This is of course not fully justified, one should at least involve the intermediate state of (2.100) in the supremum, but in practice this works quite well except an excessive numerical diffusion of waves associated to intermediate eigenvalues.

For the isentropic gas dynamics system, (2.104) gives

$$c = \max\left(|u_l| + \sqrt{p'(\rho_l)}, |u_r| + \sqrt{p'(\rho_r)}\right). \tag{2.105}$$

The Rusanov flux preserves the positiveness of density because the intermediate state in (2.100) has positive density $(\rho_l + \rho_l u_l/c)/2 + (\rho_r - \rho_r u_r/c)/2 \geq 0$ (apply Proposition 2.11), and handles data with vacuum since c does not blow up at vacuum.

2.4.3   HLL flux

A generalization of the previous solver is obtained if we take two parameters $c_1 < c_2$ (instead of $-c$ and c), and consider the relaxation system for $f = (f_1, f_2)$

$$\begin{cases} \partial_t f_1 + c_1 \partial_x f_1 = 0, \\ \partial_t f_2 + c_2 \partial_x f_2 = 0. \end{cases} \tag{2.106}$$

Then

$$\mathcal{A}(f) = (c_1 f_1, c_2 f_2), \qquad Lf = f_1 + f_2. \tag{2.107}$$

The conditions (2.68), (2.69) read $M_1(U) + M_2(U) = U$, $c_1 M_1(U) + c_2 M_2(U) = F(U)$, thus we need to take

$$M_1(U) = \frac{c_2 U - F(U)}{c_2 - c_1}, \qquad M_2(U) = \frac{-c_1 U + F(U)}{c_2 - c_1}. \tag{2.108}$$

We apply Proposition 2.13 with $\mathcal{R}$ the exact solver, which is given by

$$\mathcal{R}(x/t, f^l, f^r) = \begin{cases} (f_1^l, f_2^l) & \text{if } x/t < c_1, \\ (f_1^r, f_2^l) & \text{if } c_1 < x/t < c_2, \\ (f_1^r, f_2^r) & \text{if } c_2 < x/t, \end{cases} \tag{2.109}$$

thus we get the approximate Riemann solver

$$R(x/t, U_l, U_r) = \begin{cases} U_l & \text{if } x/t < c_1, \\ \dfrac{c_2 U_r - F(U_r)}{c_2 - c_1} + \dfrac{-c_1 U_l + F(U_l)}{c_2 - c_1} & \text{if } c_1 < x/t < c_2, \\ U_r & \text{if } c_2 < x/t. \end{cases} \tag{2.110}$$

According to (2.56), the HLL numerical flux is

$$F(U_l, U_r) = \begin{cases} F(U_l) & \text{if } 0 < c_1, \\ \dfrac{c_2 F(U_l) - c_1 F(U_r)}{c_2 - c_1} + \dfrac{c_1 c_2}{c_2 - c_1}(U_r - U_l) & \text{if } c_1 < 0 < c_2, \\ F(U_r) & \text{if } c_2 < 0. \end{cases} \tag{2.111}$$

The subcharacteristic condition, related to the invertibility of $M_1(U)$ and $M_2(U)$, is now

$$c_1 \leq \lambda_j(U) \leq c_2, \qquad (2.112)$$

and the invariant domains and entropy conditions are analyzed similarly as for the Lax–Friedrichs flux. The HLL numerical flux was introduced in [56], and was indeed the first example of approximate Riemann solver. This numerical flux is a little less diffusive than the Lax-Friedrichs flux, but has the same drawback to be too diffusive for waves corresponding to eigenvalues other than the lowest and largest ones. Again a local optimization of (2.112) leads to the choice

$$c_1 = \inf_{U = U_l, U_r} \inf_j \lambda_j(U), \qquad c_2 = \sup_{U = U_l, U_r} \sup_j \lambda_j(U). \qquad (2.113)$$

2.4.4 Suliciu relaxation system

The situation where the relaxation approach is particularly interesting is when we use a relaxation system for which it is quite easy to find the exact Riemann solution. Then we take indeed for $\mathcal{R}$ in Proposition 2.13 the exact solver. Apart from the case of a linear relaxation system, a more general situation where it is quite easy to find an exact Riemann solution is when all eigenvalues are linearly degenerate. This is what happens with the Suliciu relaxation system.

The Suliciu relaxation system is described in [93], [94], [30], [17], [26], [11], and is attached to the resolution of the isentropic gas dynamics system (1.9). It can also handle full gas dynamics, see Section 2.4.6.

A way to introduce this relaxation system is to start with a smooth solution to the isentropic system (1.9), and to derive an equation on the pressure $p(\rho)$. Developing the density equation as $\partial_t \rho + u \partial_x \rho + \rho \partial_x u = 0$, and multiplying by $p'(\rho)$, we obtain $\partial_t p(\rho) + u \partial_x p(\rho) + \rho p'(\rho) \partial_x u = 0$. Using again the density equation one gets

$$\partial_t (\rho p(\rho)) + \partial_x (\rho p(\rho) u) + \rho^2 p'(\rho) \partial_x u = 0. \qquad (2.114)$$

Then replacing $p(\rho)$ by a new variable π and $\rho^2 p'(\rho)$ by a constant c^2, we get the relaxation system

$$\begin{cases} \partial_t \rho + \partial_x (\rho u) = 0, \\ \partial_t (\rho u) + \partial_x (\rho u^2 + \pi) = 0, \\ \partial_t (\rho \pi) + \partial_x (\rho \pi u) + c^2 \partial_x u = 0. \end{cases} \qquad (2.115)$$

This system has $q = 3$ unknowns, $f = (\rho, \rho u, \rho \pi)$, for $p = 2$ unknowns ρ, ρu for the original system. Hence here $L(f_1, f_2, f_3) = (f_1, f_2)$ (observe that we make the identification (2.70)), and

$$\mathcal{A}(\rho, \rho u, \rho \pi) = \left(\rho u, \rho u^2 + \pi, \rho \pi u + c^2 u \right), \qquad (2.116)$$

where $c > 0$ is a parameter. The maxwellian equilibrium is defined here by

$$M(\rho, \rho u) = \left(\rho, \rho u, \rho p(\rho)\right), \qquad (2.117)$$

and the equations (2.68)–(2.69) are obviously satisfied. The exact resolution of the Riemann problem for (2.115) is quite simple, because it can be put in diagonal form

$$\begin{cases} \partial_t(\pi + cu) + (u + c/\rho)\partial_x(\pi + cu) = 0, \\ \partial_t(\pi - cu) + (u - c/\rho)\partial_x(\pi - cu) = 0, \\ \partial_t(1/\rho + \pi/c^2) + u\,\partial_x(1/\rho + \pi/c^2) = 0, \end{cases} \qquad (2.118)$$

and we can observe that all eigenvalues are linearly degenerate, leading to only contact discontinuities (see Example 1.9). Thus the approximate solver we get for (1.9) is simple in the sense of Section 2.3.1. The speeds and the intermediate values are given as a special case of (2.133), (2.135) when $c_l = c_r = c$.

This solver is entropy satisfying if the parameter c is chosen sufficiently large in the sense of the following subcharacteristic condition, meaning that the eigenvalues of the system to be solved lie between the eigenvalues of the relaxation system (2.115), which are $u - c/\rho$, u, $u + c/\rho$ according to (2.118).

Lemma 2.16. *If c is chosen in such a way that the Riemann solution to* (2.115) *has a density lying in some interval, $\rho(t, x) \in I$, such that $I \subset (0, \infty)$ and*

$$\forall \rho \in I, \quad \rho^2 p'(\rho) \le c^2, \qquad (2.119)$$

then the approximate Riemann solver obtained by Proposition 2.13 *preserves positiveness of density and is entropy satisfying.*

Proof. The positiveness of density follows from Proposition 2.11. For the entropy inequality, in order to apply Proposition 2.15, we have to build an entropy extension in the sense of Definition 2.14. Following [17], this is done by setting

$$\mathcal{H}(\rho, u, \pi) = \rho\, u^2/2 + \rho\, \varphi\left(1/\rho + \pi/c^2\right) + \rho\, \pi^2/2c^2, \qquad (2.120)$$

where φ is given for any $g \in J \equiv \{1/\rho + p(\rho)/c^2\}_{\rho \in I}$ by

$$\varphi(g) = \sup_{\rho \in I} \left\{e(\rho) - p(\rho)^2/2c^2 - p(\rho)\left(g - (1/\rho + p(\rho)/c^2)\right)\right\}, \qquad (2.121)$$

or equivalently by

$$\varphi\left(1/\rho + p(\rho)/c^2\right) = e(\rho) - p(\rho)^2/2c^2, \quad \rho \in I. \qquad (2.122)$$

The entropy flux is

$$\mathcal{G}(\rho, u, \pi) = \left(\mathcal{H}(\rho, u, \pi) + \pi\right) u. \qquad (2.123)$$

In order to justify these definitions, let us first prove the equivalence between (2.121) and (2.122). We notice that

$$\frac{d}{d\rho}\left(1/\rho + p(\rho)/c^2\right) = -\frac{1}{\rho^2}\left(1 - \rho^2 p'(\rho)/c^2\right) \le 0, \qquad (2.124)$$

thus $\rho \mapsto 1/\rho + p(\rho)/c^2$ is a nonincreasing function from I to J. Then, let $\chi(\rho)$ be the function between braces in (2.121). One can check that

$$\chi'(\rho) = p'(\rho)\left(1/\rho + p(\rho)/c^2 - g\right), \tag{2.125}$$

thus writing that $g = 1/\rho_g + p(\rho_g)/c^2$ for some $\rho_g \in I$ (ρ_g may be not unique), we have that $\chi'(\rho) \geq 0$ for $\rho \leq \rho_g$, and $\chi'(\rho) \leq 0$ for $\rho \geq \rho_g$. Therefore, the supremum in (2.121) is attained at $\rho = \rho_g$, which gives (2.122).

Then, we observe that according to the last equation in (2.118), $1/\rho + \pi/c^2$ remains in J, thus $\mathcal{H}$ in (2.120) is well-defined. Applying Lemma 1.4, the convexity of $\mathcal{H}$ with respect to $\rho, \rho u, \rho \pi$ is equivalent to the convexity of $u^2/2 + \varphi(1/\rho + \pi/c^2) + \pi^2/2c^2$ with respect to $(1/\rho, u, \pi)$, which is obvious since by (2.121), $\varphi(g)$ is a convex function of g. In order to prove that $\mathcal{H}$ is an entropy, we write from (2.118)

$$\begin{aligned}
&\partial_t(\pi + cu)^2 + (u + c/\rho)\partial_x(\pi + cu)^2 = 0, \\
&\partial_t(\pi - cu)^2 + (u - c/\rho)\partial_x(\pi - cu)^2 = 0, \\
&\partial_t\varphi(1/\rho + \pi/c^2) + u\,\partial_x\varphi(1/\rho + \pi/c^2) = 0,
\end{aligned} \tag{2.126}$$

thus by addition

$$(\partial_t + u\partial_x)\left(u^2/2 + \pi^2/2c^2 + \varphi(1/\rho + \pi/c^2)\right) + \frac{1}{\rho}\partial_x(\pi u) = 0, \tag{2.127}$$

which together with the first equation of (2.115) gives $\partial_t\mathcal{H} + \partial_x\mathcal{G} = 0$, proving that $\mathcal{H}$ is an entropy for (2.115), with $\mathcal{G}$ as entropy flux.

The fact that $\mathcal{H}$ and $\mathcal{G}$ are extensions of η and G in (1.25) and (1.27) is obvious since replacing π by $p(\rho)$ gives directly (2.77), (2.78).

Finally, it remains to check the minimization principle (2.79). It means here that whenever $\rho \in I$ and $1/\rho + \pi/c^2 \in J$,

$$\eta(\rho, u) \leq \mathcal{H}(\rho, u, \pi). \tag{2.128}$$

But according to (2.121),

$$\varphi(1/\rho + \pi/c^2) \geq e(\rho) - p(\rho)^2/2c^2 - p(\rho)\left(\pi - p(\rho)\right)/c^2, \tag{2.129}$$

thus

$$\begin{aligned}
\mathcal{H}(\rho, u, \pi) &\geq \rho u^2/2 + \rho\left(e(\rho) - p(\rho)^2/2c^2 - p(\rho)\left(\pi - p(\rho)\right)/c^2\right) + \rho\pi^2/2c^2 \\
&= \eta(\rho, u) + \rho(\pi - p(\rho))^2/2c^2 \\
&\geq \eta(\rho, u),
\end{aligned} \tag{2.130}$$

which gives (2.128) and concludes the Lemma. $\qquad\square$

In order to apply Lemma 2.16, we can take a value c depending on U_l, U_r which is the smallest possible to satisfy (2.119) (see Remark 2.2). An iterative procedure to compute this optimal value is proposed in [17]. However, in practice a more explicit choice is preferable, as we explain in the next section, see in particular Proposition 2.18.

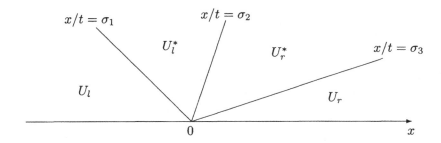

Figure 2.5: Suliciu approximate Riemann solver

2.4.5   Suliciu relaxation adapted to vacuum

The problem with the previous solver is that it cannot handle vacuum, in the sense of Section 2.3.4. Indeed since the extreme eigenvalues are $u_l - c/\rho_l$, $u_r + c/\rho_r$, we see that if one of the two densities ρ_l or ρ_r tends to 0 while the other remains finite, the propagation speed will tend to infinity, unless c tends to zero, which is not possible by (2.119) since the other density remains finite.

In order to cure this defect, we take c nonconstant in (2.115), and we choose to solve

$$\partial_t c + u\,\partial_x c = 0. \tag{2.131}$$

We see then that the whole system (2.115),(2.131) can be put in conservative form

$$\begin{cases} \partial_t \rho + \partial_x(\rho u) = 0, \\ \partial_t(\rho u) + \partial_x(\rho u^2 + \pi) = 0, \\ \partial_t(\rho \pi/c^2) + \partial_x(\rho \pi u/c^2) + \partial_x u = 0, \\ \partial_t(\rho c) + \partial_x(\rho c u) = 0. \end{cases} \tag{2.132}$$

One can check that all eigenvalues are again linearly degenerate, thus we can compute the exact solution to the Riemann problem. It has three wave speeds σ_1, σ_2, σ_3, with two intermediate states that we shall index by l^* and r^* (see Figure 2.5). We notice that $c_l^* = c_l$, $c_r^* = c_r$. Then, according to the diagonal form (2.118) and to the fact that u and π are two independent Riemann invariants for the central wave (see Section 1.5), the wave speeds are given by

$$\sigma_1 = u_l - c_l/\rho_l, \quad \sigma_2 = u_l^* = u_r^*, \quad \sigma_3 = u_r + c_r/\rho_r, \tag{2.133}$$

and the intermediate states are obtained by the relations

$$\begin{gathered} u_l^* = u_r^*, \qquad \pi_l^* = \pi_r^*, \\ (\pi + cu)_l^* = (\pi + cu)_l, \qquad (\pi - cu)_r^* = (\pi - cu)_r, \\ \left(1/\rho + \pi/c^2\right)_l^* = \left(1/\rho + \pi/c^2\right)_l, \qquad \left(1/\rho + \pi/c^2\right)_r^* = \left(1/\rho + \pi/c^2\right)_r. \end{gathered} \tag{2.134}$$

The solution is easily found to be

$$u_l^* = u_r^* = \frac{c_l u_l + c_r u_r + \pi_l - \pi_r}{c_l + c_r}, \qquad \pi_l^* = \pi_r^* = \frac{c_r \pi_l + c_l \pi_r - c_l c_r (u_r - u_l)}{c_l + c_r},$$

$$\frac{1}{\rho_l^*} = \frac{1}{\rho_l} + \frac{c_r(u_r - u_l) + \pi_l - \pi_r}{c_l(c_l + c_r)}, \qquad \frac{1}{\rho_r^*} = \frac{1}{\rho_r} + \frac{c_l(u_r - u_l) + \pi_r - \pi_l}{c_r(c_l + c_r)}.$$

$$(2.135)$$

Since we have to start with maxwellian initial data, this means that we take $\pi_l = p(\rho_l)$, $\pi_r = p(\rho_r)$. The intermediate fluxes of (2.55) are

$$F_l^* = \left(\rho_l^* u_l^*, \rho_l^* (u_l^*)^2 + \pi_l^* \right), \qquad F_r^* = \left(\rho_r^* u_r^*, \rho_r^* (u_r^*)^2 + \pi_r^* \right). \qquad (2.136)$$

The positiveness of ρ_l^* and ρ_r^* is not guaranteed *a priori* in (2.135), this is a requirement that constraints c_l, c_r to be large enough. Another requirement is that $\sigma_1 < \sigma_2 < \sigma_3$, but indeed this property follows from the previous one since one has $\sigma_2 - \sigma_1 = c_l/\rho_l^*$, $\sigma_3 - \sigma_2 = c_r/\rho_r^*$. However, as we shall see, the positiveness of ρ_l^*, ρ_r^* is less restrictive on the possible choice of c_l, c_r than the subcharacteristic condition we derive below.

Even if the system (2.132) is not strictly speaking a relaxation system, we still get a simple approximate Riemann solver for the isentropic gas dynamics system (1.9). The subcharacteristic condition of Lemma 2.16 has now to be written more precisely. Going through the analysis of [17] (the proof is provided in the more general setting of Lemma 2.20), it takes the following form.

Lemma 2.17. *With the formulas (2.135), if c_l and c_r are chosen in such a way that the densities ρ_l, ρ_r, ρ_l^*, ρ_r^* are positive and satisfy*

$$\forall \rho \in [\rho_l, \rho_l^*], \quad \rho^2 p'(\rho) \leq c_l^2,$$

$$\forall \rho \in [\rho_r, \rho_r^*], \quad \rho^2 p'(\rho) \leq c_r^2, \qquad (2.137)$$

then the approximate Riemann solver preserves positiveness of density and is entropy satisfying.

Indeed the entropy inequality then follows from the resolution of

$$\partial_t \left(\rho u^2/2 + \rho e \right) + \partial_x \left((\rho u^2/2 + \rho e + \pi)u \right) = 0, \qquad (2.138)$$

with $e_l = e(\rho_l)$, $e_r = e(\rho_r)$, the subcharacteristic conditions (2.137) ensuring the decrease at the projection step. The equation (2.138) can be combined with (2.132) to obtain

$$\partial_t(e - \pi^2/2c^2) + u\,\partial_x(e - \pi^2/2c^2) = 0, \qquad (2.139)$$

and therefore we can take for intermediate entropy fluxes

$$G_l^* = \left(\rho_l^* (u_l^*)^2/2 + \rho_l^* e_l^* + \pi_l^* \right) u_l^*, \qquad G_r^* = \left(\rho_r^* (u_r^*)^2/2 + \rho_r^* e_r^* + \pi_r^* \right) u_r^*, \quad (2.140)$$

where the intermediate values for e are

$$e_l^* = e_l - \pi_l^2/2c_l^2 + (\pi_l^*)^2/2c_l^2, \qquad e_r^* = e_r - \pi_r^2/2c_r^2 + (\pi_r^*)^2/2c_r^2. \qquad (2.141)$$

We are now going to explain how to choose the two parameters c_l, c_r, in such a way that the subcharacteristic conditions (2.137) are satisfied and that the scheme is able to treat the vacuum, in the sense of Section 2.3.4. Indeed, noticing the value of the speeds in (2.133), we need to have that c_l/ρ_l and c_r/ρ_r both remain bounded when one of the two densities ρ_l, ρ_r tends to zero, the other remaining nonzero. A possible solution is to impose the relation $c_l/\rho_l = c_r/\rho_r = a > 0$, and then to find the smallest a such that the inequalities (2.137) are satisfied. Because of the strong nonlinearity of this problem, we rather choose here a direct estimate. We make the following assumptions:

$$\forall \rho > 0, \quad \frac{d}{d\rho}\left(\rho\sqrt{p'(\rho)}\right) > 0 \qquad (2.142)$$

$$\rho\sqrt{p'(\rho)} \to \infty \quad \text{as } \rho \to \infty, \qquad (2.143)$$

$$\frac{d}{d\rho}\left(\rho\sqrt{p'(\rho)}\right) \le \alpha\sqrt{p'(\rho)}, \qquad \text{for some constant } \alpha \ge 1. \qquad (2.144)$$

These assumptions are very natural, and are satisfied for a gamma law $p(\rho) = \kappa\rho^\gamma$ with $\kappa > 0$, $\gamma \ge 1$, with the value $\alpha = (\gamma + 1)/2$. Indeed, (2.142) is equivalent to the convexity of p with respect to $1/\rho$, and according to [45] it means that both eigenvalues of the system are genuinely nonlinear.

Proposition 2.18. *Under the assumptions (2.142)–(2.144), when ρ_l, $\rho_r > 0$, define the relaxation speeds by*

$$\text{if } p_r - p_l \ge 0, \quad \begin{cases} \dfrac{c_l}{\rho_l} = \sqrt{p'(\rho_l)} + \alpha\left(\dfrac{p_r - p_l}{\rho_r\sqrt{p'(\rho_r)}} + u_l - u_r\right)_+, \\[4mm] \dfrac{c_r}{\rho_r} = \sqrt{p'(\rho_r)} + \alpha\left(\dfrac{p_l - p_r}{c_l} + u_l - u_r\right)_+, \end{cases} \qquad (2.145)$$

$$\text{if } p_r - p_l \le 0, \quad \begin{cases} \dfrac{c_r}{\rho_r} = \sqrt{p'(\rho_r)} + \alpha\left(\dfrac{p_l - p_r}{\rho_l\sqrt{p'(\rho_l)}} + u_l - u_r\right)_+, \\[4mm] \dfrac{c_l}{\rho_l} = \sqrt{p'(\rho_l)} + \alpha\left(\dfrac{p_r - p_l}{c_r} + u_l - u_r\right)_+. \end{cases} \qquad (2.146)$$

Then the intermediate densities ρ_l^*, ρ_r^* are positive and the subcharacteristic conditions (2.137) are satisfied. In particular, we obtain a positive entropy satisfying approximate Riemann solver for the isentropic gas dynamics system (1.9) that handles the vacuum.

The property to treat the vacuum is seen just by observing that there is no blow-up in (2.145)–(2.146) when one of the densities tends to 0, because of the overall positive parts. Observe also that our choice of the relaxation speeds is sharp, in the sense that when $U_l = U_r$, (2.133) gives the exact eigenvalues of $F'(U)$. This ensures the optimality of the CFL condition when U_l and U_r are not too far.

In order to prove Proposition 2.18, let us first rewrite the subcharacteristic conditions (2.137). The assumptions (2.142)–(2.143) ensure that we have an inverse function $\psi : (0, \infty) \to (0, \infty)$ such that

$$\rho\sqrt{p'(\rho)} = c \iff \rho = \psi(c). \tag{2.147}$$

Then, (2.144) means that $\psi'(c) \geq \psi(c)/\alpha c$. Writing that $\frac{d}{dc}(\psi(c)c^{-1/\alpha}) \geq 0$, we get that

$$\forall \lambda \geq 1, \quad \psi(\lambda c) \geq \lambda^{1/\alpha}\psi(c). \tag{2.148}$$

According to the monotonicity of ψ and to (2.135), the conditions (2.137) read

$$\rho_l\sqrt{p'(\rho_l)} \leq c_l, \quad \frac{1}{\rho_l} + \frac{c_r(u_r - u_l) + p_l - p_r}{c_l(c_l + c_r)} \geq \frac{1}{\psi(c_l)},$$
$$\rho_r\sqrt{p'(\rho_r)} \leq c_r, \quad \frac{1}{\rho_r} + \frac{c_l(u_r - u_l) + p_r - p_l}{c_r(c_l + c_r)} \geq \frac{1}{\psi(c_r)}. \tag{2.149}$$

Observe that these conditions include the positivity of ρ_l^* and ρ_r^*.

Lemma 2.19. *For any given $c_r > 0$, if we define*

$$\frac{c_l}{\rho_l} = \sqrt{p'(\rho_l)} + \alpha\left(\frac{p_r - p_l}{c_r} + u_l - u_r\right)_+, \tag{2.150}$$

then the two first conditions in the first line of (2.149) are met.

Proof. We have obviously $c_l \geq \rho_l\sqrt{p'(\rho_l)}$, thus the first condition is trivial. For the second, if $c_r(u_r - u_l) + p_l - p_r \geq 0$, it is obviously satisfied. Assume now that $c_r(u_r - u_l) + p_l - p_r \leq 0$, and define

$$X = \frac{p_r - p_l}{c_r} + u_l - u_r \geq 0. \tag{2.151}$$

Then $c_l = \rho_l(\sqrt{p'(\rho_l)} + \alpha X)$, thus multiplying by ρ_l, the condition reads

$$1 - \frac{c_r}{c_l + c_r}\frac{X}{\sqrt{p'(\rho_l)} + \alpha X} \geq \frac{\rho_l}{\psi(c_l)}. \tag{2.152}$$

Denoting by

$$\theta = \frac{\sqrt{p'(\rho_l)}}{\sqrt{p'(\rho_l)} + \alpha X}, \quad 1 - \theta = \frac{\alpha X}{\sqrt{p'(\rho_l)} + \alpha X}, \tag{2.153}$$

and since $c_r/(c_l + c_r) \leq 1$, it is enough to have

$$1 - \frac{1-\theta}{\alpha} - \frac{\rho_l}{\psi(\rho_l\sqrt{p'(\rho_l)}/\theta)} \geq 0. \tag{2.154}$$

Then, since $0 < \theta \leq 1$ and $\psi(\rho_l\sqrt{p'(\rho_l)}) = \rho_l$, according to (2.148) it is enough to have

$$1 - \frac{1-\theta}{\alpha} - \theta^{1/\alpha} \geq 0, \tag{2.155}$$

and this is indeed the case when $0 < \theta \leq 1$ and $\alpha \geq 1$. $\qquad\square$

Proof of Proposition 2.18. The result of Lemma 2.19 can of course be symmetrized, and for any $c_l > 0$, the value

$$\frac{c_r}{\rho_r} = \sqrt{p'(\rho_r)} + \alpha\left(\frac{p_l - p_r}{c_l} + u_l - u_r\right)_+ \tag{2.156}$$

satisfies the second line of (2.149). Consider now the choice of c_l, c_r given by (2.145)–(2.146), and assume for instance that $p_r - p_l \geq 0$. Then by the previous remark, the second line of (2.149) is satisfied. Concerning the first line, in the case $c_r(u_r - u_l) + p_l - p_r \geq 0$ it is trivial since $c_l \geq \rho_l\sqrt{p'(\rho_l)}$. Thus let us assume that $c_r(u_r - u_l) + p_l - p_r \leq 0$. Then since $c_r \geq \rho_r\sqrt{p'(\rho_r)}$ and $p_r - p_l \geq 0$, we have that $c_l \geq \widetilde{c}_l$, with

$$\frac{\widetilde{c}_l}{\rho_l} = \sqrt{p'(\rho_l)} + \alpha\left(\frac{p_r - p_l}{c_r} + u_l - u_r\right)_+. \tag{2.157}$$

But by Lemma 2.19, the couple $(\widetilde{c}_l, c_r)$ fulfills the first line of (2.149). Since $c_r(u_r - u_l) + p_l - p_r \leq 0$, this condition is monotone with respect to c_l, thus we conclude that (c_l, c_r) also satisfies this condition. $\qquad\square$

2.4.6   Suliciu relaxation/HLLC solver for full gas dynamics

2.4.6.a Reduction to an almost isentropic system

The full gas dynamics system (1.11) can be handled via a general idea introduced in [30], which consists in reversing the role of energy conservation and entropy inequality, thus considering the system

$$\begin{cases} \partial_t\rho + \partial_x(\rho u) = 0, \\ \partial_t(\rho u) + \partial_x(\rho u^2 + p) = 0, \\ \partial_t(\rho s) + \partial_x(\rho s u) = 0, \end{cases} \tag{2.158}$$

with the entropy inequality

$$\partial_t(\rho(u^2/2 + e)) + \partial_x((\rho(u^2/2 + e) + p)u) \leq 0. \tag{2.159}$$

The convexity of this entropy with respect to $(\rho, \rho u, \rho s)$ is ensured by the physically relevant condition

$$-s \text{ is a convex function of } (1/\rho, e), \qquad (2.160)$$

see [45]. Assume that we have a conservative numerical scheme for solving (2.158),

$$\rho_i^{n+1} - \rho_i + \frac{\Delta t}{\Delta x_i}\left(F^\rho_{i+1/2} - F^\rho_{i-1/2}\right) = 0,$$

$$\rho_i^{n+1} u_i^{n+1} - \rho_i u_i + \frac{\Delta t}{\Delta x_i}\left(F^{\rho u}_{i+1/2} - F^{\rho u}_{i-1/2}\right) = 0, \qquad (2.161)$$

$$\rho_i^{n+1} s_i^{n+1} - \rho_i s_i + \frac{\Delta t}{\Delta x_i}\left(F^{\rho s}_{i+1/2} - F^{\rho s}_{i-1/2}\right) = 0,$$

satisfying an entropy inequality

$$\rho_i^{n+1}((u_i^{n+1})^2/2 + e(\rho_i^{n+1}, s_i^{n+1})) - \rho_i(u_i^2/2 + e(\rho_i, s_i)) + \frac{\Delta t}{\Delta x_i}\left(F^e_{i+1/2} - F^e_{i-1/2}\right) \leq 0. \qquad (2.162)$$

We assume moreover that the scheme satisfies for any ϕ convex the inequalities

$$\rho_i^{n+1}\phi(s_i^{n+1}) - \rho_i\,\phi(s_i) + \frac{\Delta t}{\Delta x_i}\left(F^{\rho\phi(s)}_{i+1/2} - F^{\rho\phi(s)}_{i-1/2}\right) \leq 0. \qquad (2.163)$$

Then we define the scheme for the gas dynamics system (1.11) by

$$\rho_i^{n+1} - \rho_i + \frac{\Delta t}{\Delta x_i}\left(F^\rho_{i+1/2} - F^\rho_{i-1/2}\right) = 0,$$

$$\rho_i^{n+1} u_i^{n+1} - \rho_i u_i + \frac{\Delta t}{\Delta x_i}\left(F^{\rho u}_{i+1/2} - F^{\rho u}_{i-1/2}\right) = 0,$$

$$\rho_i^{n+1}((u_i^{n+1})^2/2 + e_i^{n+1}) - \rho_i(u_i^2/2 + e_i) + \frac{\Delta t}{\Delta x_i}\left(F^e_{i+1/2} - F^e_{i-1/2}\right) = 0,$$

$$(2.164)$$

where of course we take initially $s_i = s(\rho_i, e_i)$, or equivalently $e_i = e(\rho_i, s_i)$. Then by comparing the last equation in (2.164) to (2.162), we deduce that

$$e_i^{n+1} \geq e(\rho_i^{n+1}, s_i^{n+1}), \qquad (2.165)$$

where s_i^{n+1} is computed by (2.161). Therefore, $e_i^{n+1} \geq 0$, and since by (1.12) $(\frac{\partial s}{\partial e})_\rho = 1/T > 0$, this yields $s(\rho_i^{n+1}, e_i^{n+1}) \geq s_i^{n+1}$. Combining this with (2.163) we obtain for any ϕ convex and nonincreasing

$$\rho_i^{n+1}\phi(s(\rho_i^{n+1}, e_i^{n+1})) - \rho_i\,\phi(s(\rho_i, e_i)) + \frac{\Delta t}{\Delta x_i}\left(F^{\rho\phi(s)}_{i+1/2} - F^{\rho\phi(s)}_{i-1/2}\right) \leq 0, \qquad (2.166)$$

which is the desired discrete entropy inequality corresponding to (1.33) (recall Lemma 1.3).

We conclude that in order to solve the full gas dynamics system, it is enough to solve the problem (2.158)–(2.159), which is formally the same as the isentropic system (1.9), except that instead of being constant, s is transported by the flow through $\partial_t s + u\,\partial_x s = 0$.

2.4.6.b Resolution of an extended Suliciu relaxation system

The system (2.158) can be resolved with the same relaxation system (2.132), (2.138) to which we add the transport of specific entropy,

$$
\begin{cases}
\partial_t \rho + \partial_x(\rho u) = 0, \\
\partial_t(\rho u) + \partial_x(\rho u^2 + \pi) = 0, \\
\partial_t\left(\rho u^2/2 + \rho e\right) + \partial_x\left((\rho u^2/2 + \rho e + \pi)u\right) = 0, \\
\partial_t(\rho \pi/c^2) + \partial_x(\rho \pi u/c^2) + \partial_x u = 0, \\
\partial_t(\rho c) + \partial_x(\rho c u) = 0, \\
\partial_t(\rho s) + \partial_x(\rho s u) = 0.
\end{cases}
\tag{2.167}
$$

One has to take care that in (2.167), ρ, e, s, π are understood as independent variables. According to the previous paragraph, for the resolution of (2.158), the equation on ρs is taken as conservative variable and the energy e is just forgotten at time t_{n+1}, being replaced by $e(\rho^{n+1}, s^{n+1})$. On the contrary, in the case of true gas dynamics, e is kept while s is replaced by $s(\rho^{n+1}, e^{n+1})$. The last variable π is in any case replaced by the true pressure, $p(\rho^{n+1}, s^{n+1})$ or $p(\rho^{n+1}, e^{n+1})$.

The solution to the Riemann problem for (2.167) is obvious since s is not coupled, it is the same as before (2.133), (2.135), (2.141), to which we prescribe $s_l^* = s_l$, $s_r^* = s_r$. Only the initialization is modified, now

$$
\pi_l = p_l = p(\rho_l, s_l), \qquad \pi_r = p_r = p(\rho_r, s_r).
\tag{2.168}
$$

In order to get (2.162), we use the following generalization of Lemma 2.17.

Lemma 2.20. *In the Riemann problem for* (2.167) *with initially* (2.168) *and* $e_l = e(\rho_l, s_l)$, $e_r = e(\rho_r, s_r)$, *if* c_l *and* c_r *are chosen in such a way that the densities* ρ_l, ρ_r, ρ_l^*, ρ_r^* *are positive and satisfy*

$$
\begin{aligned}
\forall \rho \in [\rho_l, \rho_l^*], \quad \rho^2 \left(\frac{\partial p}{\partial \rho}\right)_s (\rho, s_l) \le c_l^2, \\
\forall \rho \in [\rho_r, \rho_r^*], \quad \rho^2 \left(\frac{\partial p}{\partial \rho}\right)_s (\rho, s_r) \le c_r^2,
\end{aligned}
\tag{2.169}
$$

then we get an approximate Riemann solver for the almost isentropic system (2.158), *that is entropy satisfying with respect to the entropy in* (2.159).

Proof. Recall that the index s in (2.169) means that the derivative is taken at s constant. Consider the three speeds $\sigma_1, \sigma_2, \sigma_3$ of (2.133), and the intermediate states for $U = (\rho, \rho u, \rho s)$ coming from the solution of the Riemann problem (2.167). In order that these values form a simple approximate solver in the sense of Section 2.3.1, we need to check the relation (2.53). But since from (2.167) we have $\partial_t U + \partial_x \Xi = 0$ with $\Xi = (\rho u, \rho u^2 + \pi, \rho s u)$, we deduce the Rankine–Hugoniot jump relations $\sigma_k(U_k - U_{k-1}) = \Xi_k - \Xi_{k-1}$. Therefore, $\sum_{k=1}^3 \sigma_k(U_k - U_{k-1}) = \Xi_3 - \Xi_0 = F(U_r) - F(U_l)$ with F the flux of (2.158), which proves the conservative relation. Notice that it gives also the fluxes of (2.55), $F_k = \Xi_k$.

In order to prove the entropy inequality (2.54), we notice that by the third equation in (2.167), $\sigma_k(E_k - E_{k-1}) = \xi_k - \xi_{k-1}$, with $E = \rho u^2/2 + \rho e$ and $\xi = (\rho u^2/2 + \rho e + \pi)u$. Recall that here e is an independent variable. We deduce with G the entropy flux of (2.159) that $G(U_r) - G(U_l) = \xi_3 - \xi_0 = \sum_{k=1}^{3} \sigma_k(E_k - E_{k-1}) = -\sigma_1 E_0 + (\sigma_1 - \sigma_2)E_1 + (\sigma_2 - \sigma_3)E_2 + \sigma_3 E_3$. But since $E_0 = \eta(U_l)$ and $E_3 = \eta(U_r)$, we conclude by comparison to (2.54) that we only need to prove that $E_1 \geq \eta(U_1)$ and $E_2 \geq \eta(U_2)$, or in other words

$$e_l^* \geq e(\rho_l^*, s_l), \quad e_r^* \geq e(\rho_r^*, s_r). \tag{2.170}$$

Notice that once this is known, one could argue here as in the previous Paragraph 2.4.6.a to conclude directly that we have an entropy approximate Riemann solver for the true gas dynamics system.

Let us now prove (2.170). We use a decomposition in elementary entropy dissipation terms along each of the three waves that was introduced in [17]. We denote now by U only the density and velocity variables, and define the derivative of the isentropic entropy in Lagrange variable at fixed s by $\mathcal{V}^s(U)$,

$$U = (\rho, \rho u), \qquad \mathcal{V}^s(U) = \left(-p(\rho, s), u\right). \tag{2.171}$$

Then we have the following decomposition along the waves,

$$
\begin{aligned}
e(\rho_r^*, s_r) - e_r^* &= D_-^r(U_r^*, \pi_r^* - c_r u_r^*) + D_+^r(U_r^*, \pi_r^* + c_r u_r^*) + D_0^r(U_r^*, U_r), \\
e(\rho_l^*, s_l) - e_l^* &= D_-^l(U_l^*, \pi_l^* - c_l u_l^*) + D_+^l(U_l^*, \pi_l^* + c_l u_l^*) + D_0^l(U_l^*, U_l),
\end{aligned}
\tag{2.172}
$$

where the upper indices l or r on the elementary dissipations D_-, D_+, D_0 mean that c, s are taken either c_l or c_r, respectively s_l or s_r, and

$$D_-(U, \Lambda) = \frac{1}{4c^2}(p - cu)^2 - \frac{1}{4c^2}\Lambda^2 - \mathcal{V}^s(U)\left(-\frac{1}{2c^2}(p - cu - \Lambda), -\frac{1}{2c}(p - cu - \Lambda)\right), \tag{2.173}$$

$$D_+(U, \Lambda) = \frac{1}{4c^2}(p + cu)^2 - \frac{1}{4c^2}\Lambda^2 - \mathcal{V}^s(U)\left(-\frac{1}{2c^2}(p + cu - \Lambda), \frac{1}{2c}(p + cu - \Lambda)\right), \tag{2.174}$$

$$D_0(U_1, U_2) = e(\rho_1, s) - \frac{p_1^2}{2c^2} - \left(e(\rho_2, s) - \frac{p_2^2}{2c^2}\right) - \mathcal{V}^s(U_1)\left(\frac{1}{\rho_1} + \frac{p_1}{c^2} - \frac{1}{\rho_2} - \frac{p_2}{c^2}, 0\right). \tag{2.175}$$

In (2.173) and (2.174), p stands for $p(\rho, s)$, while in (2.175), $p_1 = p(\rho_1, s)$, $p_2 = p(\rho_2, s)$. The identities (2.172) can be checked directly, indeed the sum of the terms in factor of $\mathcal{V}^s(U)$ vanish according to the relations of the last line in (2.134), and the sum of the terms which are not in factor of $\mathcal{V}^s(U)$ simplify with (2.141).

Now that we have (2.172), it is enough to prove that D_-, D_+, D_0 are all nonpositive. For D_- and D_+ this is trivial since using the definition of $\mathcal{V}^s(U)$ one directly simplifies

$$D_-(U,\Lambda) = -\frac{1}{4c^2}(p - cu - \Lambda)^2, \qquad D_+(U,\Lambda) = -\frac{1}{4c^2}(p + cu - \Lambda)^2. \quad (2.176)$$

For the last dissipation term D_0, we have two values ρ_1, ρ_2 but single values for c and s, thus we can use the analysis of Lemma 2.16. It ensures that whenever we have some interval I, $I \subset (0,\infty)$ satisfying (2.119), we have the equivalence of the two definitions of φ, (2.121) and (2.122). In particular, if ρ_1, $\rho_2 \in I$, one has

$$\begin{aligned}
&e(\rho_2) - p(\rho_2)^2/2c^2 \\
&= \varphi\left(1/\rho_2 + p(\rho_2)/c^2\right) \\
&\geq e(\rho_1) - p(\rho_1)^2/2c^2 - p(\rho_1)\left(1/\rho_2 + p(\rho_2)/c^2 - \left(1/\rho_1 + p(\rho_1)/c^2\right)\right),
\end{aligned} \quad (2.177)$$

thus $D_0(U_1,U_2) \leq 0$. In order to get this inequality for $D_0^r(U_r^*,U_r)$ and $D_0^l(U_l^*,U_l)$ in (2.172), we just take respectively $I = [\rho_l, \rho_l^*]$ and $I = [\rho_r, \rho_r^*]$, which satisfy (2.119) by the assumptions (2.169). $\square$

With the convention that p' stands for $(\frac{\partial p}{\partial \rho})_s$, one checks easily by using Lemma 2.20 instead of Lemma 2.17, that Proposition 2.18 is still valid with $p'(\rho, s_l)$ for the left condition and $p'(\rho, s_r)$ for the right condition. Gathering the results and using the approach explained in the previous subsection (see also Remark 2.10 in Section 2.7), we deduce the following result.

Proposition 2.21. *If (2.160) is satisfied and if the assumptions (2.142)–(2.144) hold at fixed s, the simple solver defined by the wave speeds (2.133) and intermediate states (2.135), (2.141), with (2.168) and with the choice of the relaxation speeds (2.145)–(2.146) is an approximate Riemann solver for the full gas dynamics system (1.11). It has the following properties:*
(i) it preserves the nonnegativity of ρ,
(ii) it preserves the positivity of e,
(iii) it satisfies discrete entropy inequalities,
(iv) if satisfies the maximum principle on the specific entropy.
(v) stationary contact discontinuities where $u = 0$, $p = cst$ are exactly resolved.
(vi) it handles data with vacuum.
The numerical fluxes are given by (2.136), (2.140).

Only the property (v) has not been discussed previously. It can be seen either by the formulas (2.135), (2.141), (2.133) that give for such initial data $u_l^* = u_r^* = 0$, $\pi_l^* = \pi_r^* = p$, $\rho_l^* = \rho_l$, $\rho_r^* = \rho_r$, $e_l^* = e_l$, $e_r^* = e_r$, $\sigma_2 = 0$, or either by observing more generally that a solution to the full gas dynamics system such that $u = cst$, $p = cst$, and $\partial_t \rho + u \partial_x \rho = 0$ gives a particular solution to (2.167) with $\pi = p$. In this property, the choice of c_l and c_r does not matter at all.

We can observe that the approximate Riemann solver we get can be identified with the HLLC solver that can be found in [97], [10], or the solver proposed in [32], or [37]. Here by the relaxation and subcharacteristic analysis we have provided for the first time a full proof of the entropy conditions valid for arbitrary large data and in particular for vacuum, and with explicit and sharp values of the wavespeeds, for general (convex) pressure laws.

Remark 2.3. The method works the same to treat the full gas dynamics system with transverse velocity.

2.5 Kinetic solvers

Kinetic solvers and related kinetic formulations can be found in [86]. We propose here the vector approach of [16], [17].

Definition 2.22. *A kinetic system associated to the conservation law (1.8) is an equation*

$$\partial_t f + a(\xi)\partial_x f = 0, \tag{2.178}$$

where $f(t, x, \xi) \in \mathbb{R}^p$, and $\xi \in \Xi$ is a new variable lying in a measure space Ξ with nonnegative measure $d\xi$. The link between (2.178) and (1.8) is made by the assumption that we have an equilibrium function $M(U, \xi) \in \mathbb{R}^p$, the maxwellian equilibrium, such that

$$\int_\Xi M(U, \xi)\, d\xi = U, \tag{2.179}$$

$$\int_\Xi a(\xi)M(U, \xi)\, d\xi = F(U). \tag{2.180}$$

We can see that indeed a kinetic system is a particular case of relaxation system (Definition 2.12), where we replace $\mathbb{R}^q$ by a space of infinite dimension, the space $(\mathbb{R}^p)^\Xi$ of functions of ξ with values in $\mathbb{R}^p$. The operator L is given by $Lf = \int f(\xi)d\xi$, and the flux is linear diagonal, $\mathcal{A}(f)(\xi) = a(\xi)f(\xi)$. Therefore, the notions and the results introduced in Section 2.4 remain valid. The main point is that the solution to (2.178) can be computed easily, it is given by

$$f(t, x, \xi) = f^n(x - (t - t_n)a(\xi), \xi). \tag{2.181}$$

Therefore, we take for $\mathcal{R}$ in Proposition 2.13 the exact solver, and a simple computation gives then the approximate Riemann solver

$$R(x/t, U_l, U_r) = \int_{x/t < a(\xi)} M(U_l, \xi)\, d\xi + \int_{x/t > a(\xi)} M(U_r, \xi)\, d\xi, \tag{2.182}$$

which looks like a simple solver, except that it has continuously many speeds $a(\xi)$, $\xi \in \Xi$.

Remark 2.4. We notice from (2.182) that if we have a convex invariant domain $\mathcal{U}$ which is a cone, i.e. it is stable by multiplication by positive scalars, such that $M(U, \xi) \in \mathcal{U}$ for any $U \in \mathcal{U}$ and $\xi \in \Xi$, then $R(x/t, U_l, U_r) \in \mathcal{U}$ also, and therefore according to Proposition 2.11, the solver preserves $\mathcal{U}$.

From (2.37), we get the numerical flux

$$F(U_l, U_r) = F^+(U_l) + F^-(U_r), \tag{2.183}$$

with

$$F^+(U) = \int_{a(\xi)>0} a(\xi) M(U, \xi) \, d\xi, \qquad F^-(U) = \int_{a(\xi)<0} a(\xi) M(U, \xi) \, d\xi. \tag{2.184}$$

The nonlocal approach of Section 2.4.1 is of particular interest here, it allows to justify the use of a CFL 1 condition

$$\Delta t \sup_{M(U_i, \xi) \neq 0, M(U_{i+1}, \xi) \neq 0} |a(\xi)| \leq \min(\Delta x_i, \Delta x_{i+1}), \tag{2.185}$$

just because the linear interaction of waves is fully computed in (2.181).

The entropy analysis of Section 2.4 is of course valid, and we notice that a convex entropy $\mathcal{H}$ for (2.178) and its entropy flux $\mathcal{G}$ need to be given by

$$\mathcal{H}(f) = \int_\Xi H(f(\xi), \xi) \, d\xi, \qquad \mathcal{G}(f) = \int_\Xi a(\xi) H(f(\xi), \xi) \, d\xi, \tag{2.186}$$

where $H(f, \xi)$ is a scalar function which is convex with respect to the first variable. Thus the numerical entropy flux is

$$G(U_l, U_r) = G^+(U_l) + G^-(U_r), \tag{2.187}$$

with

$$G^\pm(U) = \int_{\pm a(\xi)>0} a(\xi) H(M(U, \xi), \xi) \, d\xi. \tag{2.188}$$

According to [16], a necessary condition for $\mathcal{H}$ to be an entropy extension of η is that

$$H'(M(U, \xi), \xi) = \eta'(U). \tag{2.189}$$

This implies in particular that (2.77) and (2.78) hold (differentiate with respect to U and use (2.179), (2.180)), and that

$$G^{\pm'}(U) = \eta'(U) F^{\pm'}(U). \tag{2.190}$$

Example 2.5. The relaxation interpretation of the Lax–Friedrichs scheme (2.95)–(2.98) gives indeed a kinetic solver with $\Xi = \{-1, 1\}$, $a(-1) = -c$, $a(1) = c$, and with $d\xi$ the counting measure. Other kinetic solvers with finite sets Ξ can be found in [5].

Remark 2.6. The numerical flux associated to a kinetic method takes the form of a *flux vector splitting scheme*,

$$F(U_l, U_r) = F^+(U_l) + F^-(U_r), \qquad (2.191)$$

where F^+, F^- satisfy

$$F^+(U) + F^-(U) = F(U). \qquad (2.192)$$

Conversely, it is proved in [17] that any flux vector splitting scheme (2.191)–(2.192) can be interpreted as a kinetic method. Moreover, the entropy conditions for such a scheme can be characterized precisely, and the entropy fluxes always take the form (2.187), (2.190). A major inconvenient with flux vector splitting schemes is that for full gas dynamics they cannot resolve exactly contact discontinuities, they somehow put too much numerical viscosity.

2.5.1 Kinetic solver for isentropic gas dynamics

The classical kinetic models take the form

$$M(U, \xi) = K(\xi)\underline{M}(U, \xi), \qquad f(U, \xi) = K(\xi)\underline{f}(U, \xi), \qquad (2.193)$$

where $K(\xi) \in \mathbb{R}^p$ is a given vector, and $\underline{M}(U, \xi)$, $\underline{f}(U, \xi)$ are real. Thus the kinetic equation on f simplifies in a scalar equation

$$\partial_t \underline{f} + a(\xi)\partial_x \underline{f} = 0, \qquad (2.194)$$

and the relations (2.179),(2.180) become moment relations,

$$\int_\Xi K(\xi)\underline{M}(U, \xi) \, d\xi = U, \qquad \int_\Xi a(\xi)K(\xi)\underline{M}(U, \xi) \, d\xi = F(U). \qquad (2.195)$$

For the system of isentropic gas dynamics (1.9) and if we take

$$p(\rho) = \kappa\rho^\gamma, \qquad \kappa > 0, \ 1 < \gamma \le 3, \qquad (2.196)$$

the fundamental kinetic model is obtained as $\Xi = \mathbb{R}$ with the Lebesgue measure, $a(\xi) = \xi$, $K(\xi) = (1, \xi)$,

$$\underline{M}(U, \xi) = c_2 \left(\frac{2\gamma\kappa}{\gamma - 1}\rho^{\gamma-1} - |\xi - u|^2 \right)_+^\lambda, \qquad (2.197)$$

with

$$\lambda = \frac{1}{\gamma - 1} - \frac{1}{2}, \qquad c_2 = \left(\frac{2\gamma\kappa}{\gamma - 1} \right)^{-1/(\gamma-1)} \frac{\Gamma\left(\frac{\gamma}{\gamma-1}\right)}{\sqrt{\pi}\Gamma(\lambda + 1)}. \qquad (2.198)$$

Under a CFL 1 condition (2.185), this solver preserves positiveness of density (apply Remark 2.4), and is entropy satisfying, with here

$$H(\underline{f}, \xi) = \underline{f}|\xi|^2/2 + \frac{1}{2c_2^{1/\lambda}} \frac{\underline{f}^{1+1/\lambda}}{1 + 1/\lambda}. \qquad (2.199)$$

Thus this kinetic method naturally treats the vacuum in the sense of Section 2.3.4. However, the CFL condition (2.185) is not sharp, because when $U_i = U_{i+1} = U$, $\sup |a(\xi)| = |u| + \sqrt{\frac{2}{\gamma-1} p'(\rho)}$ and there is an overall factor $\sqrt{\frac{2}{\gamma-1}} \geq 1$ with respect to the real sound speed, that induces smaller timesteps than expected. A difficulty arises in trying to obtain explicit formulas for the integrals (2.184) involved in the numerical flux (2.183). This is not possible for all values of γ. But for $\gamma = 2$ this works with arcos functions.

2.6   VFRoe method

The VFRoe method has been introduced in [24], [38], [39]. It relies on some approximate resolution of the Riemann problem (2.34) by linearization in the spirit of the method of Roe, but it does not enter the framework of approximate Riemann solvers of Section 2.3 because the numerical flux is not deduced from taking the average of the approximate solution (2.42), but is rather defined directly as the value of the flux at the approximate interface value.

For the *nonconservative variable* version VFRoencv, the first step is, starting from the conservation law (1.8), to perform a nonlinear change of variables $Y(U)$ (with inverse $U(Y)$), to get a quasilinear system

$$\partial_t Y + B(Y)\partial_x Y = 0, \qquad (2.200)$$

with according to (1.4), $B(Y) = (dY/dU)F'(U)(dY/dU)^{-1}$. Then, as in the Roe method (2.59), we solve a linearized problem

$$\partial_t Y + B(\widehat{Y})\partial_x Y = 0, \qquad (2.201)$$

with Riemann initial data

$$Y^0(x) = \left\{ \begin{array}{ll} Y(U_l) & \text{if } x < 0, \\ Y(U_r) & \text{if } x > 0, \end{array} \right. \qquad (2.202)$$

and with

$$\widehat{Y} = \frac{Y(U_l) + Y(U_r)}{2}. \qquad (2.203)$$

Finally, the numerical flux is defined by

$$F(U_l, U_r) = F\Big(U(Y(x/t = 0, U_l, U_r))\Big), \qquad (2.204)$$

at least if 0 is not an eigenvalue of $B(\widehat{Y})$. If this is the case, $Y(x/t = 0)$ is not well defined and some other formula needs to be used, for example by defining $Y(0) = (Y(0-) + Y(0+))/2$. This numerical flux is obviously consistent.

However, it is not possible for this scheme to analyze the preservation of invariant domains and the existence of entropy inequalities. In practice the scheme can produce negative densities and violate entropy conditions (a correction is proposed in [24]). But the idea of [38] is to choose the change of variable $Y(U)$ in such a way that the scheme almost never produces negative densities for gas dynamics (1.9). This is obtained by the choice of a variable related to the Riemann invariants involved in (1.20),

$$Y = (\varphi(\rho), u), \tag{2.205}$$

where φ is defined by (1.21). According to [45], for $p(\rho) = \kappa \rho^\gamma$, the necessary and sufficient condition for the appearance of vacuum in the exact Riemann solution is

$$u_l - u_r + \varphi_l + \varphi_r \leq 0. \tag{2.206}$$

With the choice (2.205), the quasilinear formulation (2.200) of the isentropic system (1.9) becomes

$$\begin{cases} \partial_t \varphi + u \partial_x \varphi + \sqrt{p'(\rho)} \partial_x u = 0, \\ \partial_t u + u \partial_x u + \sqrt{p'(\rho)} \partial_x \varphi = 0. \end{cases} \tag{2.207}$$

Therefore, the linearized version (2.201) is

$$\begin{cases} \partial_t \varphi + \widehat{u} \partial_x \varphi + \sqrt{p'(\widehat{\rho})} \partial_x u = 0, \\ \partial_t u + \widehat{u} \partial_x u + \sqrt{p'(\widehat{\rho})} \partial_x \varphi = 0, \end{cases} \tag{2.208}$$

or in diagonal form

$$\begin{cases} \partial_t (u + \varphi) + (\widehat{u} + \sqrt{p'(\widehat{\rho})}) \partial_x (u + \varphi) = 0, \\ \partial_t (u - \varphi) + (\widehat{u} - \sqrt{p'(\widehat{\rho})}) \partial_x (u - \varphi) = 0, \end{cases} \tag{2.209}$$

with

$$\widehat{u} = \frac{u_l + u_r}{2}, \qquad \varphi(\widehat{\rho}) = \frac{\varphi(\rho_l) + \varphi(\rho_r)}{2}. \tag{2.210}$$

Solving the Riemann problem for (2.209), we get an intermediate state between the two eigenvalues, i.e. for $\lambda_1 = \widehat{u} - \sqrt{p'(\widehat{\rho})} < x/t < \lambda_2 = \widehat{u} + \sqrt{p'(\widehat{\rho})}$, which is defined by

$$u^* = \frac{1}{2}(u_l + \varphi_l + u_r - \varphi_r), \qquad \varphi^* = \frac{1}{2}(u_l + \varphi_l - u_r + \varphi_r). \tag{2.211}$$

The second formula defines a density $\rho^* > 0$ only if the assigned value is positive. Therefore, a natural way to extend the solution is to put a positive part, $\varphi^* = \frac{1}{2}(u_l + \varphi_l - u_r + \varphi_r)_+$. Then the condition of appearance of vacuum in this approximate solution is exactly the same (2.206) as for the exact solver. This property is the motivation of the choice of the variable (2.205). The numerical flux is given by (2.204), and the state at which the nonlinearity is evaluated is *left* if $\lambda_1 > 0$, *right* if $\lambda_2 < 0$, and *star* if $\lambda_1 < 0 < \lambda_2$. This numerical flux is discontinuous when one of the eigenvalues crosses 0.

2.7   Passive transport

We would like here to mention a well-known general method to solve a transport problem for the unknown ϕ

$$\partial_t(\rho\phi) + \partial_x(\rho u\phi) = 0, \tag{2.212}$$

where $\rho \geq 0$, u are assumed to be given solving

$$\partial_t\rho + \partial_x(\rho u) = 0. \tag{2.213}$$

For smooth solutions, the two equations can be combined to give

$$\partial_t\phi + u\partial_x\phi = 0. \tag{2.214}$$

This problem occurs in many fluid dynamics problems, (2.214) means that ϕ is simply passively transported with the flow. The functions ρ, u can be thought to be obtained by solving a system of equations that can involve other quantities, but we need not specify how they are obtained for what we explain here.

An important property of (2.212)-(2.213) is that it gives directly a family of entropy inequalities, because if we multiply (2.214) by $S'(\phi)$ for any function S, we get $\partial_t(S(\phi)) + u\partial_x(S(\phi)) = 0$, thus combining it with (2.213) we get $\partial_t(\rho S(\phi)) + \partial_x(\rho u S(\phi)) = 0$. Now, if S is convex, one can easily check that $\rho S(\phi)$ is a convex function of $(\rho, \rho\phi)$. Therefore, one expects for weak solutions a family of inequalities,

$$\partial_t(\rho S(\phi)) + \partial_x(\rho u S(\phi)) \leq 0, \qquad S \text{ convex.} \tag{2.215}$$

In particular, taking $S(\phi) = (\phi - k)_+$ or $S(\phi) = (k - \phi)_+$ we deduce the maximum principle

$$\inf_y \phi^0(y) \leq \phi(t, x) \leq \sup_y \phi^0(y). \tag{2.216}$$

At the numerical level, we assume that we have a discrete conservative formula for solving (2.213),

$$\rho_i^{n+1} - \rho_i + \frac{\Delta t}{\Delta x_i}\left(F_{i+1/2}^0 - F_{i-1/2}^0\right) = 0, \tag{2.217}$$

for some numerical flux $F_{i+1/2}^0$. Then the natural scheme for solving (2.212) is

$$\rho_i^{n+1}\phi_i^{n+1} - \rho_i\phi_i + \frac{\Delta t}{\Delta x_i}\left(F_{i+1/2}^\phi - F_{i-1/2}^\phi\right) = 0, \tag{2.218}$$

where

$$F_{i+1/2}^\phi = \begin{cases} F_{i+1/2}^0\,\phi_i & \text{if } F_{i+1/2}^0 \geq 0, \\ F_{i+1/2}^0\,\phi_{i+1} & \text{if } F_{i+1/2}^0 \leq 0. \end{cases} \tag{2.219}$$

This formula has been introduced in [71], and was inspired by the fact that it holds true in the exact resolution of the Riemann problem for gas dynamics equations. Denoting $x_+ = \max(0, x)$ and $x_- = \min(0, x)$, another way to write (2.219) is $F^\phi_{i+1/2} = (F^0_{i+1/2})_+ \phi_i + (F^0_{i+1/2})_- \phi_{i+1}$.

Proposition 2.23. *The upwind scheme* (2.218)–(2.219) *is consistent with* (2.212), *and under the CFL condition*

$$\rho_i - \frac{\Delta t}{\Delta x_i}(F^0_{i+1/2})_+ + \frac{\Delta t}{\Delta x_i}(F^0_{i-1/2})_- \geq 0, \tag{2.220}$$

it satisfies the discrete maximum principle and discrete entropy inequalities.

Proof. The consistency is obvious since when $\phi_i = \phi_{i+1}$, $F^\phi_{i+1/2} = F^0_{i+1/2}\phi_i$. Then, we have from (2.218)

$$
\begin{aligned}
\rho_i^{n+1}\phi_i^{n+1} = {} & \phi_i\left(\rho_i - \frac{\Delta t}{\Delta x_i}(F^0_{i+1/2})_+ + \frac{\Delta t}{\Delta x_i}(F^0_{i-1/2})_-\right) \\
& - \phi_{i+1}\frac{\Delta t}{\Delta x_i}(F^0_{i+1/2})_- + \phi_{i-1}\frac{\Delta t}{\Delta x_i}(F^0_{i-1/2})_+.
\end{aligned}
\tag{2.221}
$$

But since from (2.217)

$$
\begin{aligned}
\rho_i^{n+1} = {} & \rho_i - \frac{\Delta t}{\Delta x_i}(F^0_{i+1/2})_+ + \frac{\Delta t}{\Delta x_i}(F^0_{i-1/2})_- \\
& - \frac{\Delta t}{\Delta x_i}(F^0_{i+1/2})_- + \frac{\Delta t}{\Delta x_i}(F^0_{i-1/2})_+,
\end{aligned}
\tag{2.222}
$$

we deduce by dividing (2.221) by ρ_i^{n+1} that under the CFL condition (2.220), ϕ_i^{n+1} is a convex combination of ϕ_{i-1}, ϕ_i and ϕ_{i+1}. In particular we have the discrete form of the maximum principle,

$$\min(\phi_{i-1}, \phi_i, \phi_{i+1}) \leq \phi_i^{n+1} \leq \max(\phi_{i-1}, \phi_i, \phi_{i+1}). \tag{2.223}$$

Moreover, for any convex S, according to the Jensen inequality,

$$
\begin{aligned}
\rho_i^{n+1}S(\phi_i^{n+1}) \leq {} & S(\phi_i)\left(\rho_i - \frac{\Delta t}{\Delta x_i}(F^0_{i+1/2})_+ + \frac{\Delta t}{\Delta x_i}(F^0_{i-1/2})_-\right) \\
& - S(\phi_{i+1})\frac{\Delta t}{\Delta x_i}(F^0_{i+1/2})_- + S(\phi_{i-1})\frac{\Delta t}{\Delta x_i}(F^0_{i-1/2})_+,
\end{aligned}
\tag{2.224}
$$

therefore we have the entropy inequality

$$\rho_i^{n+1}S(\phi_i^{n+1}) - \rho_i S(\phi_i) + \frac{\Delta t}{\Delta x_i}\left(F^S_{i+1/2} - F^S_{i-1/2}\right) \leq 0, \qquad S \text{ convex}, \tag{2.225}$$

with

$$F^S_{i+1/2} = \begin{cases} F^0_{i+1/2}S(\phi_i) & \text{if } F^0_{i+1/2} \geq 0, \\ F^0_{i+1/2}S(\phi_{i+1}) & \text{if } F^0_{i+1/2} \leq 0, \end{cases} \tag{2.226}$$

which completes the proof. $\qquad\square$

Remark 2.7. Under the same CFL condition, one can check that the total variation diminishing (TVD) property holds for ϕ, by the classical Harten criterion.

The CFL condition (2.220) is indeed very natural in this context, it can be verified as follows.

Lemma 2.24. *The CFL condition* (2.220) *is automatically satisfied if the original scheme involving F^0 preserves the nonnegativity of ρ by interface, under CFL $1/2$, and if $\frac{\Delta t}{\Delta x_i}|u_i| \leq 1/2$.*

Proof. The CFL condition (2.220) can be written

$$\frac{1}{2}\left(\rho_i - \frac{2\Delta t}{\Delta x_i}((F^0_{i+1/2})_+ - \rho_i u_i)\right) + \frac{1}{2}\left(\rho_i + \frac{2\Delta t}{\Delta x_i}((F^0_{i-1/2})_- - \rho_i u_i)\right) \geq 0,$$
(2.227)

and it is enough to check that each of the two terms is nonnegative. For the first one for example, if $F^0_{i+1/2} \leq 0$, then it is obvious since $\frac{\Delta t}{\Delta x_i}|u_i| \leq 1/2$. To the contrary, if $F^0_{i+1/2} \geq 0$, this comes directly from the nonnegativity of ρ by interface and the CFL $1/2$ condition, see Definition 2.4. $\square$

Remark 2.8. The half CFL condition can be replaced by a usual CFL condition if one uses the notion of maximum principle by interface (observe that the maximum principle is a form of invariant domain property).

Remark 2.9. Another passive transport scheme can also be considered in the context of flux vector splitting schemes. Indeed, if F^0 in (2.217) comes from a FVS scheme (2.191), then we have a decomposition $F^0_{i+1/2} = F^{0+}_{i+1/2} + F^{0-}_{i+1/2}$. The condition for the scheme to preserve nonnegativity of ρ implies then that $F^{0+}_{i+1/2} \geq 0$, $F^{0-}_{i+1/2} \leq 0$. The natural scheme in the FVS spirit is then given by $F^{\phi}_{i+1/2} = F^{0+}_{i+1/2}\phi_i + F^{0-}_{i+1/2}\phi_{i+1}$. One can check that this scheme satisfies similar properties as in Proposition 2.23.

Remark 2.10. In the resolution of the Riemann problem for (2.167), s can be considered as passively transported. In fact, in this case, the numerical flux for ρs deduced from the solution of the Riemann problem coincides with the upwind flux (2.219). This is due to the fact that

$$F^0 \geq 0 \text{ if and only if } \sigma_2 = u^* \geq 0.$$
(2.228)

Indeed, when $\sigma_1 \leq 0 \leq \sigma_3$ this is obvious since $F^0 = \rho_l^* u^*$ or $F^0 = \rho_r^* u^*$ according to the sign of u^*. In the case $\sigma_1 = u_l - c_l/\rho_l \geq 0$, $F^0 = \rho_l u_l \geq c_l \geq 0$ while $u^* \geq \sigma_1 \geq 0$, and in the case $\sigma_3 = u_r + c_r/\rho_r \leq 0$, $F^0 = \rho_r u_r \leq -c_r \leq 0$ while $u^* \leq \sigma_3 \leq 0$. This proves (2.228). Then, applying Proposition 2.23 we deduce the entropy inequalities (2.163).

2.8 Second-order extension

Several methods exist to go to second-order accuracy, but we shall only describe here one general method, which has the advantage to respect invariant domains. We still consider a mesh as in (2.1), (2.2), and define

$$h = \sup_i \Delta x_i. \tag{2.229}$$

Classically, a basic ingredient in second-order schemes is a reconstruction operator.

Definition 2.25. *A second-order reconstruction is an operator which to a sequence U_i associates values $U_{i+1/2-}$, $U_{i+1/2+}$ for $i \in \mathbb{Z}$, in such a way that it is conservative,*

$$\frac{U_{i-1/2+} + U_{i+1/2-}}{2} = U_i, \tag{2.230}$$

and it is second-order in the sense that whenever for all i,

$$U_i = \frac{1}{\Delta x_i} \int_{C_i} U(x)\, dx, \tag{2.231}$$

for some smooth function $U(x)$, then

$$U_{i+1/2-} = U(x_{i+1/2}) + O(h^2), \qquad U_{i+1/2+} = U(x_{i+1/2}) + O(h^2). \tag{2.232}$$

The reconstruction is said to preserve a convex invariant domain $\mathcal{U}$ if

$$U_i \in \mathcal{U} \text{ for all } i \ \Rightarrow\ U_{i+1/2\pm} \in \mathcal{U} \text{ for all } i. \tag{2.233}$$

Once a second-order reconstruction and a first-order numerical flux $F(U_l, U_r)$ are given, we define the associated second-order scheme by

$$U_i^{n+1} - U_i^n + \frac{\Delta t}{\Delta x_i}(F_{i+1/2} - F_{i-1/2}) = 0, \tag{2.234}$$

$$F_{i+1/2} = F(U_{i+1/2-}^n, U_{i+1/2+}^n). \tag{2.235}$$

We can justify the second-order accuracy in the weak sense, as in Proposition 2.2.

Proposition 2.26. *Assume that the numerical flux is Lipschitz continuous and consistent. If for all i,*

$$U_i^n = \frac{1}{\Delta x_i} \int_{C_i} U(t_n, x)\, dx, \tag{2.236}$$

for some smooth solution $U(t,x)$ to (1.8), then U_i^{n+1} defined by (2.234)–(2.235) satisfies for all i

$$U_i^{n+1} = \frac{1}{\Delta x_i} \int_{C_i} U(t_{n+1}, x)\, dx + \Delta t \left(\frac{1}{\Delta x_i}(\mathcal{F}_{i+1/2} - \mathcal{F}_{i-1/2}) \right), \tag{2.237}$$

where

$$\mathcal{F}_{i+1/2} = O(\Delta t) + O(h^2), \tag{2.238}$$

as Δt and h tend to 0.

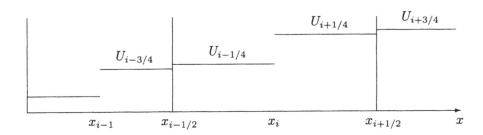

Figure 2.6: Half mesh values

Proof. It is similar to the one of Proposition 2.2. We have that (2.237) holds with $\mathcal{F}_{i+1/2} = \underline{F}_{i+1/2} - F_{i+1/2}$, and $\underline{F}_{i+1/2}$ is the exact flux (2.13). Then, since the numerical flux is Lipschitz continuous and consistent, and by (2.232),

$$F_{i+1/2} = F(U_{i+1/2-}^n, U_{i+1/2+}^n) = F(U(t_n, x_{i+1/2})) + O(h^2). \qquad (2.239)$$

Since $\underline{F}_{i+1/2} = F(U(t_n, x_{i+1/2})) + O(\Delta t)$, we obtain the result. $\qquad\qquad\square$

Remark 2.11. As explained after Proposition 2.2, the convergence is in the weak sense. Consequently, even if the rate of convergence is h^2 in the weak sense, if we measure the error in strong L^1 norm, it might converge to 0 with a slower rate. When the solution is not smooth, there is an even bigger loss of accuracy.

It is extremely difficult to obtain second-order schemes that verify an entropy inequality. Only the preservation of invariant domains can be analyzed.

Proposition 2.27. *If under a CFL condition the numerical flux preserves a convex invariant domain $\mathcal{U}$ in the sense of Definition 2.3, and if the reconstruction also preserves this invariant domain, then under the half original CFL condition, the second-order scheme also preserves this invariant domain.*

Proof. Assume that $U_i^n \in \mathcal{U}$ are given. Since the reconstruction is $\mathcal{U}$ preserving we have also that $U_{i+1/2\pm}^n \in \mathcal{U}$. Consider then a submesh with cells half of the original ones, with nodes $\cdots < x_{i-1} < x_{i-1/2} < x_i < x_{i+1/2} < \ldots$, and define discrete data on this mesh $U_{i-3/4} = U_{i-1/2-}$, $U_{i-1/4} = U_{i-1/2+}$, $U_{i+1/4} = U_{i+1/2-}, \ldots$ corresponding respectively to the cells $]x_{i-1}, x_{i-1/2}[$, $]x_{i-1/2}, x_i[$, $]x_i, x_{i+1/2}[, \ldots$ (see Figure 2.6). By applying the first-order scheme to these data, we get new values

$$\begin{aligned}
U_{i-1/4}^{n+1} &= U_{i-1/4} - \frac{2\Delta t}{\Delta x_i}\left(F_i - F_{i-1/2}\right), \\
U_{i+1/4}^{n+1} &= U_{i+1/4} - \frac{2\Delta t}{\Delta x_i}\left(F_{i+1/2} - F_i\right),
\end{aligned} \qquad (2.240)$$

with
$$F_i = F(U_{i-1/4}, U_{i+1/4}), \qquad F_{i+1/2} = F(U_{i+1/4}, U_{i+3/4}). \qquad (2.241)$$

We notice that with the definition of $U_{i+1/4}$ and $U_{i+3/4}$, the value of $F_{i+1/2}$ in (2.241) coincides with the one of (2.235) (at least if the numerical flux does not depend explicitly on the size of the mesh). But since the numerical flux preserves $\mathcal{U}$, we deduce that $U_{i-1/4}^{n+1}$, $U_{i+1/4}^{n+1} \in \mathcal{U}$ under the half original CFL condition because the half mesh has half size of the original one. Finally (and only now), we invoke the conservativity (2.230) that tells that $(U_{i-1/4} + U_{i+1/4})/2 = U_i$. By summing the two equations in (2.240) and comparing to (2.234), we conclude that $U_i^{n+1} = (U_{i-1/4}^{n+1} + U_{i+1/4}^{n+1})/2 \in \mathcal{U}$ by convexity. $\qquad\square$

The second-order reconstruction must also be nonoscillatory in some sense, but we refer to [33] or [44] for such notions. We only give here the most well known examples of reconstructions.

Example 2.12. In the case of a scalar function $U \in \mathbb{R}$, the second-order minmod reconstruction is defined as follows,

$$U_{i-1/2+} = U_i - \frac{\Delta x_i}{2} DU_i, \qquad U_{i+1/2-} = U_i + \frac{\Delta x_i}{2} DU_i, \qquad (2.242)$$

with

$$DU_i = \text{minmod}\left(\frac{U_i - U_{i-1}}{(\Delta x_{i-1} + \Delta x_i)/2}, \frac{U_{i+1} - U_i}{(\Delta x_i + \Delta x_{i+1})/2}\right), \qquad (2.243)$$

and

$$\text{minmod}(x, y) = \begin{cases} \min(x, y) & \text{if } x, y \geq 0, \\ \max(x, y) & \text{if } x, y \leq 0, \\ 0 & \text{otherwise.} \end{cases} \qquad (2.244)$$

Example 2.13. As commented in Remark 2.11, a loss of accuracy can come from the weak consistency formulation. A remedy for this is to ensure that not only $U_{i+1/2\pm}$ are second-order accurate, but also the discrete derivative DU_i. This is what does the second-order ENO (essentially non oscillatory) reconstruction. For a uniform mesh, it is defined as follows for a scalar sequence U_i,

$$U_{i-1/2+} = U_i - \frac{\Delta x}{2} D_{eno} U_i, \qquad U_{i+1/2-} = U_i + \frac{\Delta x}{2} D_{eno} U_i, \qquad (2.245)$$

where

$$D_{eno} U_i = \text{minmod}\left(\frac{U_i - U_{i-1}}{\Delta x} + \frac{\Delta x}{2} D^2 U_{i-1/2}, \frac{U_{i+1} - U_i}{\Delta x} - \frac{\Delta x}{2} D^2 U_{i+1/2}\right), \qquad (2.246)$$

$$D^2 U_{i+1/2} = \text{minmod}\left(\frac{U_{i+1} - 2U_i + U_{i-1}}{\Delta x^2}, \frac{U_{i+2} - 2U_{i+1} + U_i}{\Delta x^2}\right). \qquad (2.247)$$

A counterpart of this increased accuracy is the loss of the maximum principle, which is often needed, for example for positiveness of density. A possible way to obtain this is to consider the modified ENO reconstruction, defined still by (2.245) but with slopes

$$D_{enom}U_i = \text{minmod} \left(D_{eno}U_i, 2D_{mm}U_i \right),\qquad (2.248)$$

where $D_{mm}U_i$ is the minmod slope of (2.243) (the coefficient 2 can also be lowered down a bit for security). Then the maximum principle is recovered, at the price of loosing the second-order accuracy of the slope close to local extrema.

Example 2.14. For the isentropic gas dynamics system (1.9), a second-order reconstruction can be performed for $U_i = (\rho_i, \rho_i u_i)$ as follows. Let us denote $U_{i+1/2\pm} \equiv (\rho_{i+1/2\pm}, \rho_{i+1/2\pm} u_{i+1/2\pm})$ the reconstructed values. Then the conservation constraint (2.230) reads

$$\frac{\rho_{i-1/2+} + \rho_{i+1/2-}}{2} = \rho_i,$$
$$\frac{\rho_{i-1/2+} u_{i-1/2+} + \rho_{i+1/2-} u_{i+1/2-}}{2} = \rho_i u_i. \qquad (2.249)$$

It is easily seen to be equivalent to the representation

$$\rho_{i-1/2+} = \rho_i - \frac{\Delta x_i}{2} D\rho_i, \qquad \rho_{i+1/2-} = \rho_i + \frac{\Delta x_i}{2} D\rho_i,$$
$$u_{i-1/2+} = u_i - \frac{\rho_{i+1/2-}}{\rho_i} \frac{\Delta x_i}{2} Du_i, \qquad (2.250)$$
$$u_{i+1/2-} = u_i + \frac{\rho_{i-1/2+}}{\rho_i} \frac{\Delta x_i}{2} Du_i,$$

for some slopes $D\rho_i$, Du_i. Moreover, the second-order accuracy (2.232) means that $D\rho_i$, Du_i have to be consistent with $d\rho/dx$ and du/dx respectively. Thus we can take for $D\rho_i$ and Du_i the minmod reconstruction (2.243), where we put the values ρ_i or u_i respectively in the right-hand side. A variant is possible where we use the D_{enom} slope (2.248) for $D\rho_i$ and the D_{eno} slope (2.246) for Du_i.

Example 2.15. Let us consider now the full gas dynamics system (1.11), for which the conservative variable is

$$U \equiv \left(\rho, \rho u, \rho u^2/2 + \rho e \right). \qquad (2.251)$$

Given the values U_i, we would like to find second-order accurate values $U_{i+1/2\pm}$ such that

$$\frac{U_{i-1/2+} + U_{i+1/2-}}{2} = U_i. \qquad (2.252)$$

Knowing that the U_i have positive densities and positive internal energies

$$\rho_i > 0, \qquad e_i > 0, \qquad (2.253)$$

we require the same properties for the reconstructed values,

$$\rho_{i+1/2\pm} > 0, \qquad e_{i+1/2\pm} > 0. \tag{2.254}$$

We could also require the maximum principle on the specific entropy, but this would lead to very complicated formulas, and in practice the stability seems to be not affected by this property.

The first step of the construction is to write a parametrization of all possible values $U_{i-1/2+}$, $U_{i+1/2-}$ satisfying (2.252). An algebraic computation gives that they must be of the form

$$\rho_{i-1/2+} = \rho_i - \frac{\Delta x_i}{2} D\rho_i, \qquad \rho_{i+1/2-} = \rho_i + \frac{\Delta x_i}{2} D\rho_i \tag{2.255}$$

for some slope $D\rho_i$,

$$u_{i-1/2+} = u_i - \frac{\rho_{i+1/2-}}{\rho_i} \frac{\Delta x_i}{2} Du_i,$$

$$u_{i+1/2-} = u_i + \frac{\rho_{i-1/2+}}{\rho_i} \frac{\Delta x_i}{2} Du_i \tag{2.256}$$

for some slope Du_i, and

$$\rho_{i-1/2+} e_{i-1/2+} = \rho_i \tilde{e}_i - \frac{\Delta x_i}{2} D(\rho e)_i,$$

$$\rho_{i+1/2-} e_{i+1/2-} = \rho_i \tilde{e}_i + \frac{\Delta x_i}{2} D(\rho e)_i \tag{2.257}$$

for some slope $D(\rho e)_i$, with

$$\tilde{e}_i = e_i - \frac{\rho_{i-1/2+} \rho_{i+1/2-}}{\rho_i^2} \frac{\Delta x_i^2}{8} Du_i^2. \tag{2.258}$$

We observe then that the second-order accuracy means that $D\rho_i$, Du_i, $D(\rho e)_i$ must be consistent with the derivatives of ρ, u, ρe respectively. Concerning the positivity conditions (2.254), they can be expressed as

$$\frac{\Delta x_i}{2}|D\rho_i| < \rho_i, \qquad \frac{\Delta x_i^2}{8} Du_i^2 < \frac{\rho_i}{\rho_{i-1/2+} \rho_{i+1/2-}} \left(\rho_i e_i - \frac{\Delta x_i}{2}|D(\rho e)_i| \right). \tag{2.259}$$

Therefore, the computation of the slopes can be done as follows: we first compute $D\rho_i$, Du_i, $D(\rho e)_i$ by the minmod slope formula (2.243) where the U_i are replaced by ρ_i, u_i, $\rho_i e_i$ respectively, and then we eventually diminish the absolute value of Du_i so as to satisfy the second inequality in (2.259). This computation gives consistent values of $D\rho_i$, Du_i, $D(\rho e)_i$, thus the reconstruction is second-order accurate. It has the good property to give constant $u_{i+1/2\pm}$ and $(\rho e)_{i+1/2\pm}$ if u_i and $\rho_i e_i$ are constant, so that it is especially adapted to the computation of contact discontinuities for a gamma pressure law where p and ρe are proportional (recall that contact discontinuities are characterized by the fact that u and p do not jump).

2.8.1   Second-order accuracy in time

The second-order accuracy in time is usually recovered by the Heun method, which reads as follows. The second-order method in x defined by (2.234)–(2.235) can be written as

$$U^{n+1} = U^n + \Delta t\, \Phi(U^n), \qquad (2.260)$$

where $U = (U_i)_{i\in\mathbb{Z}}$, and Φ is a nonlinear operator depending on the mesh. Then the second-order scheme in time and space is

$$
\begin{aligned}
\widetilde{U}^{n+1} &= U^n + \Delta t\, \Phi(U^n), \\
\widetilde{U}^{n+2} &= \widetilde{U}^{n+1} + \Delta t\, \Phi(\widetilde{U}^{n+1}), \\
U^{n+1} &= \frac{U^n + \widetilde{U}^{n+2}}{2}.
\end{aligned}
\qquad (2.261)
$$

It is easy to see that if the numerical flux does not depend explicitly on Δt, this procedure gives a fully second-order scheme in the sense that we get $O(\Delta t^2) + O(h^2)$ in (2.238). The invariant domains are also preserved because of the average in (2.261), without any further limitation on the CFL.

2.9   Numerical tests

Since many tests can be found in the literature on all the schemes discussed, we only assess here the vacuum treatment, that has been especially analyzed in the previous sections.

Test 1: Rarefaction into vacuum for isentropic gas dynamics

We consider the isentropic system (1.9), with pressure law $p(\rho) = \kappa\rho^\gamma$, $\gamma > 1$, $\kappa > 0$. The initial data is the one of a Riemann problem, $U^0(x) = U_l$ for $x < x_0$, $U^0(x) = U_r$ for $x > x_0$, with vacuum on the left $U_l = 0$. The exact solution is then given by

$$
\begin{cases}
u(t,x) = \dfrac{2}{\gamma+1}\left(\max\left(\dfrac{x-x_0}{t}, -\dfrac{2}{\gamma-1}\sqrt{p'(\rho_r)} \right) - \sqrt{p'(\rho_r)} \right)_{-}, \\[4mm]
\rho(t,x) = \rho_r \left(1 + \dfrac{\gamma-1}{2}\dfrac{u(t,x)}{\sqrt{p'(\rho_r)}} \right)_{+}^{2/(\gamma-1)}.
\end{cases}
\qquad (2.262)
$$

Note that the value of u where $\rho = 0$ is irrelevant in this exact solution. However, it could influence the result in the numerical methods. The parameters are chosen as

$$\gamma = 2, \qquad \kappa = 1, \qquad (2.263)$$

$$\rho_l = 0,\; u_l = 0, \qquad \rho_r = 1,\; u_r = 0. \qquad (2.264)$$

Cells	HLL		Suliciu		Kinetic		VFRoe	
50	3.19E-2	/	2.83E-2	/	2.70E-2	/	2.49E-2	/
100	2.03E-2	0.65	1.83E-2	0.63	1.85E-2	0.55	1.73E-2	0.53
200	1.25E-2	0.70	1.16E-2	0.66	1.19E-2	0.64	1.14E-2	0.60
400	7.51E-3	0.74	7.18E-3	0.69	7.42E-3	0.68	7.21E-3	0.66
800	4.47E-3	0.75	4.39E-3	0.71	4.49E-3	0.72	4.41E-3	0.71

Table 2.1: L^1 error and numerical order of accuracy for Test 1, first-order

The runs use $x_0 = 0.5$, x lies in $[0,1]$, and the final time is $t = 0.15$. Each of the methods tested really handles vanishing densities, without need of putting small positive values, because even if in the analysis positive densities were considered, the numerical fluxes nevertheless extend continuously until the vacuum.

Table 2.1 shows the L^1 error $\sum_i \Delta x_i(|\rho_i - \rho(x_i)| + |\rho_i u_i - \rho(x_i)u(x_i)|)$ at final time, for four methods that handle vacuum: the HLL solver of Section 2.4.3, the Suliciu solver of Section 2.4.5, the kinetic solver of Section 2.5.1 and the VFRoe solver of Section 2.6. Table 2.1 shows also the numerical order of accuracy, computed by comparison between two runs with different mesh sizes. It is by definition the real number α such that the error can be written $C\Delta x^\alpha$. The CFL number used is 1 for all methods except VFRoe for which we take 0.99 to prevent negative densities (note however the overall restriction by a factor $\sqrt{2}$ inherent in the kinetic method). Recall that only the situation of Figure 2.4 could make fail the value 1 of the CFL for a scheme that is stable interface by interface, indeed this never occurs in practice. To give an idea of the CPU we provide the number of timesteps used for 100 cells, respectively 33, 35, 37, 35. For the VFRoe method we use the entropy fix of [24] otherwise it gives a wrong solution. The density and velocity profiles are plotted on Figures 2.7, 2.8, 2.9 for 30 cells.

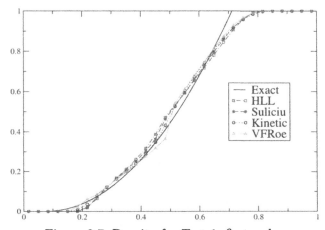

Figure 2.7: Density for Test 1, first-order

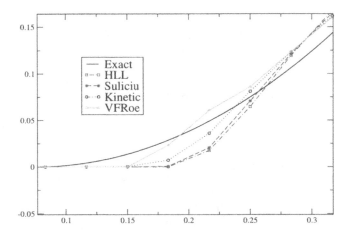

Figure 2.8: Zoom of density for Test 1, first-order

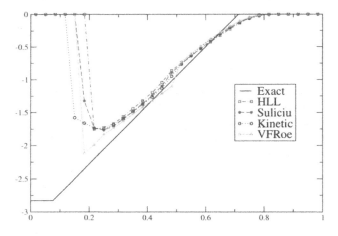

Figure 2.9: Velocity for Test 1, first-order

On Tables 2.2 and 2.3 are reported the same diagnostics for the second-order in time and space method described in Section 2.8, Example 2.14, with either the minmod limiter or the ENOm limiter of Examples 2.12, 2.13. The CFL condition is taken half of the one of the first-order method. Figures 2.10, 2.11 show a zoom of the density for 30 cells.

The results of the four methods are quite similar. We see that the presence of vacuum induces extremely low numerical orders of accuracy. The second-order methods with minmod or ENOm slope limiters give the same rate of convergence, the latter improving only by an approximate factor 0.6. On Figures 2.8 and 2.9 we can see that the density and velocity are systematically under-estimated on

Cells	HLL		Suliciu		Kinetic		VFRoe	
50	2.16E-2	/	2.03E-2	/	1.84E-2	/	1.83E-2	/
100	1.11E-2	0.96	1.05E-2	0.95	9.88E-3	0.90	9.88E-3	0.89
200	5.60E-3	0.99	5.29E-3	0.99	5.17E-3	0.93	5.29E-3	0.90
400	2.85E-3	0.97	2.69E-3	0.98	2.63E-3	0.98	2.70E-3	0.97
800	1.44E-3	0.98	1.37E-3	0.97	1.34E-3	0.97	1.37E-3	0.98

Table 2.2: L^1 error and numerical order of accuracy for Test 1, second-order minmod

Cells	HLL		Suliciu		Kinetic		VFRoe	
50	1.27E-2	/	1.12E-2	/	8.76E-3	/	8.47E-3	/
100	6.68E-3	0.93	5.84E-3	0.94	4.62E-3	0.92	4.25E-3	0.99
200	3.47E-3	0.94	3.05E-3	0.94	2.46E-3	0.91	2.17E-3	0.97
400	1.77E-3	0.97	1.57E-3	0.96	1.27E-3	0.95	1.09E-3	0.99
800	8.99E-4	0.98	7.97E-4	0.98	6.52E-4	0.96	5.49E-4	0.98

Table 2.3: L^1 error and numerical order of accuracy for Test 1, second-order ENO

the front. The kinetic solver behaves a little better than the Suliciu solver in this respect. On Figures 2.10 and 2.11 we observe the same phenomenon at second-order, but the ENO reconstruction greatly improves this behavior.

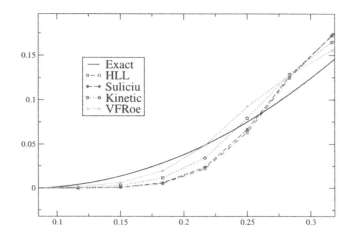

Figure 2.10: Zoom of density for Test 1, second-order minmod

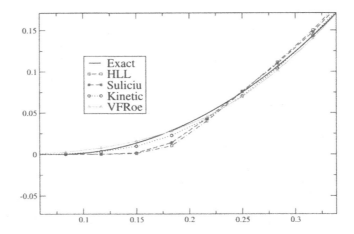

Figure 2.11: Zoom of density for Test 1, second-order ENO

Test 2: Rarefaction into vacuum for full gas dynamics

We consider now the full gas dynamics system (1.11), for a perfect polytropic gas $p(\rho, e) = (\gamma - 1)\rho e$ with $\gamma > 1$. The initial data is that of a Riemann problem with vacuum on the left $U_l = 0$. Defining $\kappa = p_r/\rho_r^\gamma$, the exact solution is then the same as in the isentropic case (2.262), with $e(t, x) = \kappa \rho(t, x)^{\gamma-1}/(\gamma - 1)$. The choice of the parameters corresponds to Test 1, $\gamma = 2$ and

$$\rho_l = 0, \; u_l = 0, \; e_l = 10^{-15}, \qquad \rho_r = 1, \; u_r = 0, \; e_r = 1/(\gamma - 1). \qquad (2.265)$$

We use again $x_0 = 0.5$, x lies in $[0, 1]$, and the final time is $t = 0.15$. We only consider here the Suliciu/HLLC solver of Section 2.4.6 (Proposition 2.21). Table 2.4 shows the L^1 error at final time and the numerical order of accuracy, for first-order with CFL number 1, and for second-order in time-space with CFL 1/2, with the reconstruction of Example 2.15 and respectively the minmod and ENOm slope evaluations. On Figures 2.12 and 2.13 are plotted the density and velocity at first-order for 30 points. They are compared to the corresponding results obtained in Test 1 by the isentropic algorithm, which differs in the fact that energy is dissipated instead of being conserved, and the specific entropy is conserved instead of being dissipated. Figure 2.14 shows the internal energy for first-order and second-order runs for 30 points.

We can see on Table 2.4 that again the ENO reconstruction gives just a factor of improvement with respect to the minmod reconstruction. The convergence rates at first or second order are even lower than in the isentropic case. However, we observe on Figures 2.12 and 2.13 that the front behaves much better than in the isentropic case.

Cells	First-order		Second-order minmod		Second-order ENO	
50	4.90E-2	/	3.60E-2	/	2.57E-2	/
100	3.56E-2	0.46	2.07E-2	0.80	1.54E-2	0.74
200	2.47E-2	0.53	1.19E-2	0.79	8.96E-3	0.78
400	1.65E-2	0.58	6.66E-3	0.84	4.88E-3	0.88
800	1.07E-2	0.62	3.69E-3	0.85	2.69E-3	0.86

Table 2.4: L^1 error and numerical order of accuracy for Test 2

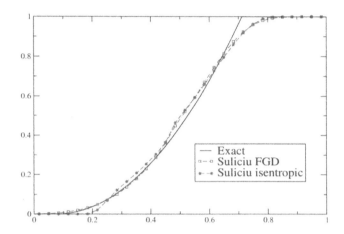

Figure 2.12: Density for Test 2, first-order

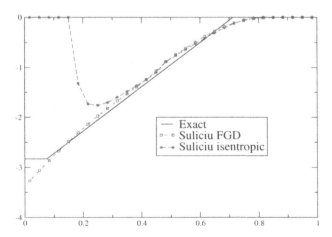

Figure 2.13: Velocity for Test 2, first-order

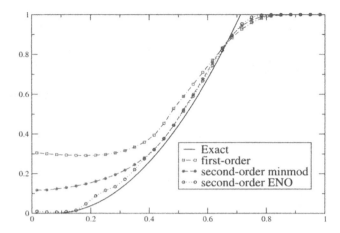

Figure 2.14: Internal energy for Test 2

Chapter 3

Source terms

The aim of this section is to describe special features arising when the system of conservation laws (1.8) is completed with a source term. We shall consider in Chapters 3 and 4 systems of the form

$$\partial_t U + \partial_x(F(U, Z)) + B(U, Z)Z_x = 0, \tag{3.1}$$

where $U(t, x) \in \mathbb{R}^p$ is the unknown, $Z(x) \in \mathbb{R}^r$ is a smooth vector valued function, and $Z_x = \partial_x Z$. The nonlinearities are supposed to be smooth also, $F(U, Z) \in \mathbb{R}^p$, and $B(U, Z)$ is a $p \times r$ matrix.

The case when Z is scalar ($r = 1$) is already interesting, and it covers the special choice $Z(x) = x$, which writes if we take $F = F(U)$, $B = B(U)$,

$$\partial_t U + \partial_x(F(U)) = -B(U). \tag{3.2}$$

This system (3.2) is the most simple system of conservation laws with source, and it has already the interesting structure which comes from the competition between the differential term and the right-hand side during the time evolution. In particular, the *steady states* are the solutions $U(x)$ which are independent of time, and hence solve $\partial_x(F(U)) = -B(U)$. These solutions play an important role because they are usually obtained as limits when time tends to infinity of the general solutions of (3.2).

The advantage of the formulation (3.1) is that this problem can be interpreted as a quasilinear system in the variable $\widetilde{U} = (U, Z)$,

$$\begin{cases} \partial_t U + \partial_x(F(U, Z)) + B(U, Z)Z_x = 0, \\ \partial_t Z = 0. \end{cases} \tag{3.3}$$

However, this system has a nonconservative term $B(U, Z)Z_x$, which is not well defined if Z is discontinuous. The problem of giving a sense to the solution to quasilinear systems which are not in conservative form is extremely difficult in general, and we refer to [72] for this question. In our case, since our main goal is to solve (3.1) for smooth Z, this question will be of minor concern, even if we nevertheless consider (3.3) for discontinuous Z, in particular when for numerical purpose we replace Z by a piecewise constant function. Indeed when we shall consider a solution to (3.3) with discontinuous Z, this means a solution in any reasonable sense, the only condition being that the generalized product $B(U, Z)Z_x$ should coincide with the usual product if Z or $B(U, Z)$ is continuous. Examples of possible definitions are provided in [47], [48]. We have to mention that however, the solutions of (3.3) are almost well-defined, see below. The formulation (3.1) also includes when $B = 0$ the problem of discontinuous flux function, if for example Z takes only two values.

We shall always assume that the system is hyperbolic with respect to U, which means that $F_U(U, Z)$ is diagonalizable. Critical points will play a special role in the whole system (3.3).

Definition 3.1. *A critical (or resonant) point for* (3.3) *is a point* (U, Z) *such that* $F_U(U, Z)$ *is not invertible.*

Lemma 3.2. *The quasilinear system* (3.3) *is hyperbolic at every noncritical point.*

Proof. The matrix of the system (in the sense of Section 1.1) is

$$A(U, Z) = \begin{pmatrix} F_U & F_Z + B \\ 0 & 0 \end{pmatrix}. \qquad (3.4)$$

Thus the eigenvalues of $A(U, Z)$ are those of F_U, to which we adjoin the value 0. At a noncritical point, F_U is diagonalizable and does not have 0 as eigenvalue, thus obviously $A(U, Z)$ is diagonalizable.   □

The occurrence of resonant points prevents from having a smooth dependence of the eigenvalues and eigenvectors, as was stated in Section 1.1. However, we can observe that outside of the resonance, we have a smooth dependence, and the eigenvalue 0 is obviously linearly degenerate. In particular, 0-contact discontinuities are well-defined, even if the system is not in conservative form (see Section 1.5). Since here the only nonconservative term involves Z_x, and Z is stationary, the only difficulty could come from stationary discontinuities, which are 0-contact discontinuities. Therefore, finally, the solutions are indeed well-defined out of the resonance, at least for piecewise smooth solutions. Notice that contact discontinuities associated to the vanishing eigenvalue are in particular steady states, which explains the critical role played by these solutions. In practice, the Riemann invariants associated to the 0-wave can be found each time it is possible to write conservative equations, because by the proof of Lemma 1.7, the associated flux is a Riemann invariant.

Nevertheless, our aim here is not to give a detailed description of the consequences of the existence of nonconservative products and resonant points to the resolution of the Cauchy problem for (3.1). We rather restrict here to the notions we shall use in the context of numerical methods. Concerning theoretical results, the reader is referred in particular to [81], [82], [57], [58], [59], [84], [99], [1], [2], [42], [73], [54] for sources, and to [96], [41], [69], [70], [64] for discontinuous fluxes.

3.1   Invariant domains and entropy

The notions of invariant domains and entropy inequalities are available for the quasilinear system (3.3) and we do not repeat here the definitions of Sections 1.3

and 1.4. Of course here an entropy $\tilde{\eta}$ and its entropy flux $\tilde{G}$ are both functions of (U, Z). Here $\tilde{\eta}$ need only be convex with respect to U, because the equation on Z is linearly degenerate.

We can define also the notion of *partial entropy* $\eta(U, Z)$ and *partial entropy flux* $G(U, Z)$ that satisfy $G_U = \eta_U F_U$, and for which we have

$$\partial_t \left(\eta(U, Z) \right) + \partial_x \left(G(U, Z) \right) + Q(U, Z) Z_x \leq 0, \tag{3.5}$$

with $Q = \eta_U(F_Z + B) - G_Z$. This notion is enough if we only consider smooth $Z(x)$. Again η need only be convex with respect to U.

3.2 Saint Venant system

The main example of a system with source, resonance and nontrivial steady states is the Saint Venant system for shallow water with topography. This system is naturally under the form (3.1). We denote here $\rho(t, x) \geq 0$ the water height in analogy with the isentropic gas dynamics system (1.9), and $u(t, x)$ the velocity. Then the system reads

$$\begin{cases} \partial_t \rho + \partial_x(\rho u) = 0, \\ \partial_t(\rho u) + \partial_x(\rho u^2 + p(\rho)) + \rho g z_x = 0, \end{cases} \tag{3.6}$$

where $g > 0$ is the gravitational constant and $z(x)$ is the topography. We shall denote

$$Z = gz, \tag{3.7}$$

hence the system has the form (3.3) with $U = (\rho, \rho u)$,

$$F(U, Z) = F(U) = (F^0(U), F^1(U)) = (\rho u, \rho u^2 + p(\rho)), \tag{3.8}$$

$$B(U, Z) = B(U) = (B^0(U), B^1(U)) = (0, \rho). \tag{3.9}$$

We assume the hyperbolicity with respect to U, $p'(\rho) > 0$. The physically relevant case is indeed $p(\rho) = g\rho^2/2$, but we shall deal with the generalized case $p(\rho)$. Then the critical points are defined by $u = \pm\sqrt{p'(\rho)}$.

Steady states

In order to obtain the steady states, we subtract u times the first equation in (3.6) to the second, and divide the result by ρ. We get

$$\partial_t u + \partial_x \left(u^2/2 + e(\rho) + p(\rho)/\rho + Z \right) = 0, \tag{3.10}$$

where the internal energy is still defined by $e'(\rho) = p(\rho)/\rho^2$. Therefore, the steady states are exactly the functions $\rho(x)$, $u(x)$ satisfying

$$\begin{cases} \rho u = cst, \\ \dfrac{u^2}{2} + e(\rho) + \dfrac{p(\rho)}{\rho} + Z = cst. \end{cases} \tag{3.11}$$

Indeed, out of the resonance, the two expressions on the left-hand side are two independent 0-Riemann invariants (see the comments above). Between these steady states, some play an important role, the steady state *at rest* for which the first constant is 0, or equivalently

$$\begin{cases} u = 0, \\ e(\rho) + \dfrac{p(\rho)}{\rho} + Z = cst. \end{cases} \tag{3.12}$$

In the physically relevant case $p(\rho) = g\rho^2/2$, the second equation simplifies to $\rho + z = cst$.

Entropy

An entropy can be obtained as follows. We multiply the first equation in (3.6) by $u^2/2 + e(\rho) + p(\rho)/\rho$, we multiply (3.10) by ρu, and add the results. This gives

$$\partial_t \eta + \partial_x G + \rho u Z_x \leq 0, \tag{3.13}$$

where η and G are defined by (1.25) and (1.27). The inequality stands here just because of discontinuous solutions, as usual. Thus η is a partial entropy. Next, we add to (3.13) the first equation in (3.6) multiplied by Z, and since $\partial_t Z = 0$, this yields

$$\partial_t(\eta + \rho Z) + \partial_x(G + \rho u Z) \leq 0. \tag{3.14}$$

Thus we have the entropy and entropy flux

$$\widetilde{\eta} = \eta + \rho Z, \qquad \widetilde{G} = G + \rho u Z. \tag{3.15}$$

A remark is that we can indeed take the inequality (3.13) as entropy inequality instead of (3.14), because here it is really equivalent to (3.14). Indeed the nonconservative product $\rho u Z_x$ is well-defined, because Z has only stationary discontinuities, and ρu is the flux in the first equation of (3.6), and thus does not have any jump through stationary curves.

Other specific properties

The Saint Venant system has other specific properties that are worthwhile to state. The first is that, as in gas dynamics, the density needs to remain nonnegative. Also the total amount of water need to be preserved, which means that the first equation is conservative. Another property is that the system (3.6) is invariant under translations in Z, in the sense that adding a constant to Z does not modify the equations. The inequality (3.13) has the advantage to have this property also, in contrast with (3.14).

Chapter 4

Nonconservative schemes

The numerical treatment of sources can be performed very classically by the fractional step method and with an ODE solver [77]. However, the big defect of this approach is that it gives very inaccurate results when U is close to a steady state. Therefore, methods taking both terms into account in a coherent way need to be used. Several methods have been proposed [49], [52], [53], [76], [12], [13], [100], [40], [7], [61], [25], [48], [38], [6]. Even if our main concern is for sources, we give also for completion a few references for the case of discontinuous fluxes, [79], [80], [98], [9], [90], and for the case of source with small parameter, [60], [62], [85], [68], [50].

We shall retain here the formulation by interface source [65], which is quite general and flexible. It includes most of the known methods. It handles data Z_i attached to each cell, instead of interface values used for example in [61].

A main idea is to treat the system (3.1) as a quasilinear system (3.3), so that the differential term and source term are both resolved at the same level. Then, when considering piecewise constant data and solving the Riemann problem, we are faced to solutions involving some kind of nonconservative product, as was commented in Chapter 3. However, this incursion is only temporary since at the end we wish to converge to a continuous profile Z.

In order to solve (3.3), we consider discrete data $\widetilde{U}_i^n = (U_i^n, Z_i)$ over a mesh, similarly as in Chapter 2. We consider first-order three point nonconservative schemes that take the form

$$U_i^{n+1} - U_i^n + \frac{\Delta t}{\Delta x_i}(F_{i+1/2-} - F_{i-1/2+}) = 0, \tag{4.1}$$

with

$$F_{i+1/2-} = F_l(\widetilde{U}_i^n, \widetilde{U}_{i+1}^n), \qquad F_{i+1/2+} = F_r(\widetilde{U}_i^n, \widetilde{U}_{i+1}^n). \tag{4.2}$$

The functions

$$F_l(\widetilde{U}_l, \widetilde{U}_r) \equiv F_l(U_l, U_r, Z_l, Z_r), \qquad F_r(\widetilde{U}_l, \widetilde{U}_r) \equiv F_r(U_l, U_r, Z_l, Z_r) \tag{4.3}$$

are the left and right numerical fluxes. Of course, we need not write any formula for Z because we always use the trivial equation $Z_i^{n+1} = Z_i^n$.

4.1   Well-balancing

A main feature that is desirable for the scheme (4.1)–(4.2) is that it preserves some
discrete steady states, approximating the exact ones defined as smooth functions
$(U(x), Z(x))$ satisfying

$$\partial_x(F(U,Z)) + B(U,Z)Z_x = 0. \tag{4.4}$$

These discrete steady states are discrete sequences $(U_i, Z_i)_{i \in \mathbb{Z}}$ that satisfy an approximation of (4.4), under the form of a nonlinear relation at each interface,
linking $U_i, U_{i+1}, Z_i, Z_{i+1}$,

$$\mathcal{D}(U_i, U_{i+1}, Z_i, Z_{i+1}) = 0. \tag{4.5}$$

We shall often write this relation only locally, as $\mathcal{D}(U_l, U_r, Z_l, Z_r) = 0$. These
discrete steady states can be defined in various ways.

Example 4.1. For a scalar law $U \in \mathbb{R}$, and if $F(U,Z) = F(U)$, $B(U,Z) = B(U) >$
0, we can define $D(U)$ by

$$D'(U) = \frac{F'(U)}{B(U)}. \tag{4.6}$$

Then (4.4) becomes

$$\partial_x(D(U) + Z) = 0. \tag{4.7}$$

Therefore, we can take for discrete steady states the relation

$$D(U_l) + Z_l = D(U_r) + Z_r. \tag{4.8}$$

Example 4.2. For the Saint Venant system, since the continuous steady states are
those solving (3.11), we can take for discrete steady states the relations

$$\begin{cases} \rho_l u_l = \rho_r u_r, \\ \dfrac{u_l^2}{2} + e(\rho_l) + \dfrac{p(\rho_l)}{\rho_l} + Z_l = \dfrac{u_r^2}{2} + e(\rho_r) + \dfrac{p(\rho_r)}{\rho_r} + Z_r. \end{cases} \tag{4.9}$$

In particular, the discrete steady states at rest are those for which

$$\begin{cases} u_l = u_r = 0, \\ e(\rho_l) + \dfrac{p(\rho_l)}{\rho_l} + Z_l = e(\rho_r) + \dfrac{p(\rho_r)}{\rho_r} + Z_r. \end{cases} \tag{4.10}$$

Example 4.3. If we assume that $F = F(U)$ and $B = B(U)$ do not depend on Z in
the system (3.3), we can take for discrete steady states the relation

$$F(U_r) - F(U_l) + B(U_l, U_r, \Delta Z)\Delta Z = 0, \tag{4.11}$$

where $\Delta Z = Z_r - Z_l$, and $B(U_l, U_r, \Delta Z)$ is any consistent discretization of $B(U)$.

Once some discrete steady states are selected, we define the well-balanced schemes as follows.

Definition 4.1. *The scheme (4.1)–(4.2) is well-balanced relatively to some discrete steady state if one has for this steady state*

$$F_l(U_l, U_r, Z_l, Z_r) = F(U_l, Z_l), \qquad F_r(U_l, U_r, Z_l, Z_r) = F(U_r, Z_r). \qquad (4.12)$$

According to (4.1)–(4.2), this property guarantees obviously that if at time t_n we start with a steady state sequence (U_i), then it remains unchanged at the next time level.

4.2 Consistency

One can find quite strange the fact that the source does not appear explicitly in (4.1). This formula is however consistent, as we shall now justify, generalizing [88], [65].

Definition 4.2. *We say that the scheme (4.1)–(4.2) is consistent with (3.1) if the numerical fluxes satisfy the consistency with the exact flux*

$$F_l(U, U, Z, Z) = F_r(U, U, Z, Z) = F(U, Z) \text{ for any } (U, Z) \in \mathbb{R}^p \times \mathbb{R}^r, \qquad (4.13)$$

and the asymptotic conservativity/consistency with the source

$$F_r(U_l, U_r, Z_l, Z_r) - F_l(U_l, U_r, Z_l, Z_r) = -B(U, Z)(Z_r - Z_l) + o(Z_r - Z_l), \qquad (4.14)$$

as U_l, $U_r \to U$ and Z_l, $Z_r \to Z$.

We shall always assume the minimal regularity F_l, F_r continuous and $\delta F = F_r - F_l$ of class C^1.

A more general assumption than (4.14) is

$$\begin{aligned} &F_r(U_l, U_r, Z_l, Z_r) - F_l(U_l, U_r, Z_l, Z_r) \\ &= -B(U, Z)(Z_r - Z_l) + o(|U_l - U| + |U_r - U| + |Z_l - Z| + |Z_r - Z|), \end{aligned} \qquad (4.15)$$

as U_l, $U_r \to U$ and Z_l, $Z_r \to Z$. This means equivalently, since $\delta F \in C^1$, that $\delta F(U, U, Z, Z) = 0$ (but this is contained in (4.13)), and

$$\begin{aligned} \partial_1 \delta F(U, U, Z, Z) &= 0, & \partial_2 \delta F(U, U, Z, Z) &= 0, \\ \partial_3 \delta F(U, U, Z, Z) &= B(U, Z), & \partial_4 \delta F(U, U, Z, Z) &= -B(U, Z). \end{aligned} \qquad (4.16)$$

Another natural assumption is that the differential term becomes conservative for smooth Z, which can be stated as

$$F_l(U_l, U_r, Z, Z) = F_r(U_l, U_r, Z, Z). \qquad (4.17)$$

We can then observe that (4.14) means exactly the conjunction of (4.15) and (4.17). Indeed, if (4.14) holds, then obviously (4.15) and (4.17) hold. Conversely, if (4.15) and (4.17) hold, then

$$
\begin{aligned}
\delta F(U_l, U_r, Z_l, Z_r) &= \delta F(U_l, U_r, Z_l, Z_r) - \delta F(U_l, U_r, Z_l, Z_l) \\
&= \int_0^1 \partial_4 \delta F\Big(U_l, U_r, Z_l, (1-\theta)Z_l + \theta Z_r\Big) d\theta \, (Z_r - Z_l),
\end{aligned}
\tag{4.18}
$$

and as U_l, $U_r \to U$ and Z_l, $Z_r \to Z$, the integral tends to $\partial_4 \delta F(U, U, Z, Z) = -B(U, Z)$ by (4.16), thus we deduce that (4.14) holds.

Remark 4.4. In Definition 4.2, condition (4.14) only requires a property when Z_l and Z_r are asymptotically close, thus it only concerns continuous functions $Z(x)$, and there is no information about any consistency with the source for discontinuous $Z(x)$. However, discontinuities of U are properly handled because of the property (4.17) that ensures conservativity for continuous Z and therefore suitable Rankine–Hugoniot conditions.

The previous definition is justified by the following estimate, which is again formulated in the weak sense.

Proposition 4.3. *Assume that for all i,*

$$
U_i^n = \frac{1}{\Delta x_i} \int_{C_i} U(t_n, x) \, dx, \qquad Z_i = Z(x_i) + O(\Delta x_i),
\tag{4.19}
$$

for some smooth solution $(U(t, x), Z(x))$ to (3.3), and define U_i^{n+1} by (4.1)–(4.2). If the scheme is consistent in the generalized sense (4.13), (4.15), then for all i,

$$
U_i^{n+1} = \frac{1}{\Delta x_i} \int_{C_i} U(t_{n+1}, x) \, dx + \Delta t \left(\frac{1}{\Delta x_i} (\mathcal{F}_{i+1/2} - \mathcal{F}_{i-1/2}) + \mathcal{E}_i \right),
\tag{4.20}
$$

where

$$
\mathcal{F}_{i+1/2} \to 0, \qquad \mathcal{E}_i \to 0,
\tag{4.21}
$$

as Δt and $h = \sup_i \Delta x_i$ tend to 0.

Proof. Integrate the equation (3.1) satisfied by $U(t, x)$ with respect to t and x over $]t_n, t_{n+1}[\times C_i$, and divide the result by Δx_i. We obtain

$$
\frac{1}{\Delta x_i} \int_{C_i} U(t_{n+1}, x) \, dx - \frac{1}{\Delta x_i} \int_{C_i} U(t_n, x) \, dx + \frac{\Delta t}{\Delta x_i} (\underline{F}_{i+1/2} - \underline{F}_{i-1/2}) + \Delta t \underline{E}_i = 0,
\tag{4.22}
$$

where $\underline{F}_{i+1/2}$ is the exact flux

$$
\underline{F}_{i+1/2} = \frac{1}{\Delta t} \int_{t_n}^{t_{n+1}} F\Big(U(t, x_{i+1/2}), Z(x_{i+1/2})\Big) dt,
\tag{4.23}
$$

and $\underline{E}_i$ is the exact source

$$\underline{E}_i = \frac{1}{\Delta t \Delta x_i} \int_{t_n}^{t_{n+1}} \int_{C_i} B\Big(U(t,x), Z(x)\Big) Z_x(x) \, dt dx. \tag{4.24}$$

Then, define

$$Z_{i+1/2} = Z(x_{i+1/2}), \qquad B_{i+1/2} = B\Big(U(t_n, x_{i+1/2}), Z(x_{i+1/2})\Big), \tag{4.25}$$

$$B_i = B\Big(U(t_n, x_i), Z(x_i)\Big), \tag{4.26}$$

and the mean flux

$$F_{i+1/2} = \frac{\Delta x_{i+1} F_{i+1/2-} + \Delta x_i F_{i+1/2+}}{\Delta x_i + \Delta x_{i+1}} + B_{i+1/2} \left(\frac{\Delta x_{i+1} Z_i + \Delta x_i Z_{i+1}}{\Delta x_i + \Delta x_{i+1}} - Z_{i+1/2} \right).$$

$$\tag{4.27}$$

The equation (4.1) can be rewritten

$$U_i^{n+1} - U_i^n + \frac{\Delta t}{\Delta x_i}(F_{i+1/2} - F_{i-1/2}) + \Delta t E_i = 0, \tag{4.28}$$

with

$$\begin{aligned}
E_i &= \frac{1}{\Delta x_i}(F_{i+1/2-} - F_{i+1/2} + F_{i-1/2} - F_{i-1/2+}) \\
&= \frac{F_{i+1/2-} - F_{i+1/2+}}{\Delta x_i + \Delta x_{i+1}} + \frac{F_{i-1/2-} - F_{i-1/2+}}{\Delta x_{i-1} + \Delta x_i} \\
&\quad - \frac{B_{i+1/2}}{\Delta x_i} \left(\frac{\Delta x_{i+1} Z_i + \Delta x_i Z_{i+1}}{\Delta x_i + \Delta x_{i+1}} - Z_{i+1/2} \right) \\
&\quad + \frac{B_{i-1/2}}{\Delta x_i} \left(\frac{\Delta x_i Z_{i-1} + \Delta x_{i-1} Z_i}{\Delta x_{i-1} + \Delta x_i} - Z_{i-1/2} \right).
\end{aligned} \tag{4.29}$$

Therefore, by subtracting (4.22) to (4.28), we get (4.20) with

$$\mathcal{F}_{i+1/2} = \underline{F}_{i+1/2} - F_{i+1/2}, \qquad \mathcal{E}_i = \underline{E}_i - E_i. \tag{4.30}$$

We observe that

$$\underline{F}_{i+1/2} = F\Big(U(t_n, x_{i+1/2}), Z(x_{i+1/2})\Big) + O(\Delta t), \tag{4.31}$$

and that from (4.13), since the numerical fluxes are continuous,

$$F_{i+1/2} = F\Big(U(t_n, x_{i+1/2}), Z(x_{i+1/2})\Big) + o(1), \tag{4.32}$$

which gives $\mathcal{F}_{i+1/2} = O(\Delta t) + o(1)$. Then, (4.15) gives

$$\begin{aligned}
F_{i+1/2+} - F_{i+1/2-} &= -B_{i+1/2}(Z_{i+1} - Z_i) + o(\Delta x_i + \Delta x_{i+1}) \\
&= -B_i(Z_{i+1} - Z_i) + o(\Delta x_i + \Delta x_{i+1}),
\end{aligned} \tag{4.33}$$

and similarly

$$F_{i-1/2+} - F_{i-1/2-} = -B_{i-1/2}(Z_i - Z_{i-1}) + o(\Delta x_{i-1} + \Delta x_i)$$
$$= -B_i(Z_i - Z_{i-1}) + o(\Delta x_{i-1} + \Delta x_i). \tag{4.34}$$

Therefore, putting (4.33) and (4.34) in (4.29) and using that $B_{i+1/2} = B_i + O(\Delta x_i)$, $B_{i-1/2} = B_i + O(\Delta x_i)$, it yields

$$
\begin{aligned}
E_i &= B_i \frac{Z_{i+1} - Z_i}{\Delta x_i + \Delta x_{i+1}} + B_i \frac{Z_i - Z_{i-1}}{\Delta x_{i-1} + \Delta x_i} \\
&\quad - \frac{B_i}{\Delta x_i} \left(\frac{\Delta x_{i+1} Z_i + \Delta x_i Z_{i+1}}{\Delta x_i + \Delta x_{i+1}} - Z_{i+1/2} \right) \\
&\quad + \frac{B_i}{\Delta x_i} \left(\frac{\Delta x_i Z_{i-1} + \Delta x_{i-1} Z_i}{\Delta x_{i-1} + \Delta x_i} - Z_{i-1/2} \right) + o(1) \\
&= B_i \frac{Z_{i+1/2} - Z_{i-1/2}}{\Delta x_i} + o(1) \\
&= B_i Z_x(x_i) + o(1).
\end{aligned}
\tag{4.35}
$$

We conclude with (4.24) that $\mathcal{E}_i = o(1)$.   □

Remark 4.5. The smoothness of the numerical fluxes is more involved in Proposition 4.3 than in the classical conservative case. Here in order to get a better rate than (4.21) we can assume for example that $F_l, F_r \in C^{0,\alpha}$ and $\delta F = F_r - F_l \in C^{1,\alpha}$ with $0 < \alpha \le 1$. Then we get $O(h^\alpha)$ in (4.32) thus $\mathcal{F}_{i+1/2} = O(\Delta t) + O(h^\alpha)$, and (4.15) holds with an error in $O(|U_l - U|^{1+\alpha} + |U_r - U|^{1+\alpha} + |Z_l - Z|^{1+\alpha} + |Z_r - Z|^{1+\alpha})$, and this gives that $\mathcal{E}_i = O(\Delta t) + O(h^\alpha)$.

4.3 Stability

Stability can be analyzed, as in the conservative case, via invariant domains and entropy. The definitions and properties of the scheme with respect to the preservation of invariant domains that do not involve the variable Z in their definition are the same as in Section 2.2.1, except that we need to replace the numerical flux $F(U_l, U_r)$ in (2.17) by $F_l(U_l, U_r, Z_l, Z_r)$ and $F_r(U_l, U_r, Z_l, Z_r)$ respectively. Thus we do not recopy the statements and proofs, which are identical. We only write explicitly the results concerning entropy inequalities.

Definition 4.4. *We say that the scheme* (4.1)–(4.2) *satisfies a discrete entropy inequality associated to the convex entropy $\widetilde{\eta}$ for* (3.3), *if there exists a numerical entropy flux function $\widetilde{G}(U_l, U_r, Z_l, Z_r)$ which is consistent with the exact entropy flux (in the sense that $\widetilde{G}(U, U, Z, Z) = \widetilde{G}(U, Z)$), such that, under some CFL condition, the discrete values computed by* (4.1)–(4.2) *automatically satisfy*

$$\widetilde{\eta}(U_i^{n+1}, Z_i) - \widetilde{\eta}(U_i^n, Z_i) + \frac{\Delta t}{\Delta x_i}(\widetilde{G}_{i+1/2} - \widetilde{G}_{i-1/2}) \le 0, \tag{4.36}$$

with

$$\widetilde{G}_{i+1/2} = \widetilde{G}(U_i^n, U_{i+1}^n, Z_i, Z_{i+1}). \tag{4.37}$$

Definition 4.5. *We say that the numerical fluxes F_l, F_r satisfy an interface entropy inequality associated to the convex entropy $\widetilde{\eta}$ for (3.3), if there exists a numerical entropy flux function $\widetilde{G}(U_l, U_r, Z_l, Z_r)$ which is consistent with the exact entropy flux (in the sense that $\widetilde{G}(U, U, Z, Z) = \widetilde{G}(U, Z)$), such that for some $\sigma_l(U_l, U_r, Z_l, Z_r) < 0 < \sigma_r(U_l, U_r, Z_l, Z_r)$,*

$$\widetilde{G}(U_r, Z_r) + \sigma_r \left[\widetilde{\eta}\left(U_r + \frac{F_r(U_l, U_r, Z_l, Z_r) - F(U_r, Z_r)}{\sigma_r}, Z_r \right) - \widetilde{\eta}(U_r, Z_r) \right]$$
$$\leq \widetilde{G}(U_l, U_r, Z_l, Z_r), \tag{4.38}$$

$$\widetilde{G}(U_l, U_r, Z_l, Z_r)$$
$$\leq \widetilde{G}(U_l, Z_l) + \sigma_l \left[\widetilde{\eta}\left(U_l + \frac{F_l(U_l, U_r, Z_l, Z_r) - F(U_l, Z_l)}{\sigma_l}, Z_l \right) - \widetilde{\eta}(U_l, Z_l) \right]. \tag{4.39}$$

Lemma 4.6. *The left-hand side of (4.38) and the right-hand side of (4.39) are nonincreasing functions of σ_r and σ_l respectively. In particular, for (4.38) and (4.39) to hold it is necessary that the inequalities obtained when $\sigma_r \to \infty$ and $\sigma_l \to -\infty$ (semi-discrete limit) hold,*

$$\widetilde{G}(U_r, Z_r) + \widetilde{\eta}'(U_r, Z_r)(F_r(U_l, U_r, Z_l, Z_r) - F(U_r, Z_r)) \leq \widetilde{G}(U_l, U_r, Z_l, Z_r), \tag{4.40}$$

$$\widetilde{G}(U_l, U_r, Z_l, Z_r) \leq \widetilde{G}(U_l, Z_l) + \widetilde{\eta}'(U_l, Z_l)(F_l(U_l, U_r, Z_l, Z_r) - F(U_l, Z_l)). \tag{4.41}$$

The proof is the same as in Lemma 2.8, noticing that $\widetilde{\eta}$ is convex with respect to the first variable. In (4.40), (4.41), $\widetilde{\eta}'$ denotes the derivative of $\widetilde{\eta}$ with respect to the first argument. Remark 2.1 is also valid here, and we have the same result as in Proposition 2.9, with same proof.

Proposition 4.7. (i) *If the scheme is entropy satisfying (Definition 4.4), then its numerical fluxes are entropy satisfying by interface (Definition 4.5), with $\sigma_l = -\Delta x_i/\Delta t$, $\sigma_r = \Delta x_{i+1}/\Delta t$.*
(ii) *If the numerical fluxes are entropy satisfying by interface (Definition 4.5), then the scheme is entropy satisfying (Definition 4.4), under the half CFL condition $|\sigma_l(U_i, U_{i+1}, Z_i, Z_{i+1})|\Delta t \leq \Delta x_i/2$, $\sigma_r(U_{i-1}, U_i, Z_{i-1}, Z_i)\Delta t \leq \Delta x_i/2$.*

4.4 Required properties for Saint Venant schemes

In order to be more explicit, let us apply the notions previously introduced to the particular case of Saint Venant system. This will give all the properties that an ideal scheme should satisfy.

Following Section 3.2, the natural schemes for solving (3.6) should depend only on $\Delta Z = Z_r - Z_l$, and not separately on Z_l and Z_r, in order to preserve the translation invariance with respect to Z. Thus a generic scheme for Saint Venant system reads, with $U_i^n = (\rho_i^n, \rho_i^n u_i^n)$,

$$U_i^{n+1} - U_i^n + \frac{\Delta t}{\Delta x_i}(F_{i+1/2-} - F_{i-1/2+}) = 0, \tag{4.42}$$

with

$$F_{i+1/2-} = F_l(U_i^n, U_{i+1}^n, \Delta Z_{i+1/2}), \qquad F_{i+1/2+} = F_r(U_i^n, U_{i+1}^n, \Delta Z_{i+1/2}), \tag{4.43}$$

$$\Delta Z_{i+1/2} = Z_{i+1} - Z_i. \tag{4.44}$$

Conservativity

Denoting $F_l = (F_l^0, F_l^1)$ and $F_r = (F_r^0, F_r^1)$, the conservativity of the water height reads

$$F_l^0 = F_r^0 \equiv F^0. \tag{4.45}$$

Consistency

Taking (4.45) into account, the consistency (4.13)–(4.14) becomes

$$\begin{cases} F^0(U, U, 0) = \rho u, \\ F_l^1(U, U, 0) = F_r^1(U, U, 0) = \rho u^2 + p(\rho), \end{cases} \tag{4.46}$$

$$F_r^1(U_l, U_r, \Delta Z) - F_l^1(U_l, U_r, \Delta Z) = -\rho \Delta Z + o(\Delta Z), \tag{4.47}$$

as U_l, $U_r \to U$ and $\Delta Z \to 0$.

Well-balancing

A natural requirement is the well-balancing property (4.12) only for the discrete steady states at rest (4.10). This gives for any ρ_l, $\rho_r \geq 0$,

$$\begin{aligned} F_l\Big((\rho_l, 0), (\rho_r, 0), e(\rho_l) + \frac{p(\rho_l)}{\rho_l} - e(\rho_r) - \frac{p(\rho_r)}{\rho_r}\Big) &= \Big(0, p(\rho_l)\Big), \\ F_r\Big((\rho_l, 0), (\rho_r, 0), e(\rho_l) + \frac{p(\rho_l)}{\rho_l} - e(\rho_r) - \frac{p(\rho_r)}{\rho_r}\Big) &= \Big(0, p(\rho_r)\Big). \end{aligned} \tag{4.48}$$

Vacuum

As in the conservative gas dynamics system, an ideal scheme for Saint Venant system should keep water height ρ nonnegative under some CFL condition that does not blow up at vacuum, as explained in Section 2.3.4.

Entropy inequality

An entropy inequality may also be required. Since (3.14) is not translation invariant in Z, we rather discretize (3.13) as

$$\eta(U_i^{n+1}) - \eta(U_i^n) + \frac{\Delta t}{\Delta x_i}\left(G_{i+1/2-} - G_{i-1/2+}\right) \leq 0, \qquad (4.49)$$

$$
\begin{aligned}
G_{i+1/2-} &= G_{i+1/2} + \frac{1}{2}F^0_{i+1/2}\Delta Z_{i+1/2}, \\
G_{i+1/2+} &= G_{i+1/2} - \frac{1}{2}F^0_{i+1/2}\Delta Z_{i+1/2},
\end{aligned}
\qquad (4.50)
$$

with $\Delta Z_{i+1/2} = Z_{i+1} - Z_i$, and $G_{i+1/2} = G(U_i^n, U_{i+1}^n, \Delta Z_{i+1/2})$ is a mean numerical entropy flux consistent with the exact flux,

$$G(U,U,0) = \left(\rho u^2/2 + \rho e(\rho) + p(\rho)\right) u. \qquad (4.51)$$

We notice that adding Z_i times the first equation in (4.42) to (4.49), we get

$$\eta(U_i^{n+1}) + \rho_i^{n+1}Z_i - \eta(U_i^n) - \rho_i^n Z_i + \frac{\Delta t}{\Delta x_i}\left(\widetilde{G}_{i+1/2} - \widetilde{G}_{i-1/2}\right) \leq 0, \qquad (4.52)$$

with

$$\widetilde{G}_{i+1/2} = G_{i+1/2-} + F^0_{i+1/2}Z_i = G_{i+1/2+} + F^0_{i+1/2}Z_{i+1} = G_{i+1/2} + F^0_{i+1/2}\frac{Z_i + Z_{i+1}}{2}, \qquad (4.53)$$

thus we recover the formulation (4.36)–(4.37) corresponding to (3.14).

4.5 Explicitly well-balanced schemes

By explicitly well-balanced schemes we mean schemes that are defined via the resolution of the discrete steady states. Such schemes were first derived for scalar laws, where an exact resolution of the Riemann problem for (3.3) directly involves such discrete steady states [49], [52], [53], [14], [88], [46], [74].

We shall not discuss the general case for equation (3.1), but only consider an example of such scheme when $F = F(U)$, $B = B(U)$ do not depend on Z. Let us define the discrete steady states, as in Example 4.3, by the relation

$$F(U_r) - F(U_l) + B(U_l, U_r, \Delta Z)\Delta Z = 0, \qquad (4.54)$$

where $\Delta Z = Z_r - Z_l$, and $B(U_l, U_r, \Delta Z)$ is any consistent discretization of $B(U)$. Assume that F has no critical points. Then one should be able to find some unique values U_l^{*r} and U_r^{*l} satisfying

$$
\begin{aligned}
F(U_l^{*r}) - F(U_l) + B(U_l, U_r, \Delta Z)\Delta Z &= 0, \\
F(U_r) - F(U_r^{*l}) + B(U_l, U_r, \Delta Z)\Delta Z &= 0.
\end{aligned}
\qquad (4.55)
$$

We define the numerical fluxes by

$$F_l(U_l, U_r, \Delta Z) = \mathcal{F}(U_l, U_r^{*l}), \qquad F_r(U_l, U_r, \Delta Z) = \mathcal{F}(U_l^{*r}, U_r), \qquad (4.56)$$

where $\mathcal{F}(U_l, U_r)$ is any consistent C^1 numerical flux for solving the equation without source.

Proposition 4.8. *The scheme* (4.55)–(4.56) *is well-balanced and consistent.*

Proof. If we have a discrete steady state in the sense of (4.54), then obviously $U_l^{*r} = U_r$ and $U_r^{*l} = U_l$. Therefore,

$$F_l(U_l, U_r, \Delta Z) = \mathcal{F}(U_l, U_l) = F(U_l), \qquad (4.57)$$

and

$$F_r(U_l, U_r, \Delta Z) = \mathcal{F}(U_r, U_r) = F(U_r), \qquad (4.58)$$

which gives the well-balanced identities (4.12).

Next, the first consistency identity (4.13) is obvious. Let us prove (4.14). We have

$$\begin{aligned}
&F_r(U_l, U_r, \Delta Z) - F_l(U_l, U_r, \Delta Z) \\
&= \mathcal{F}(U_l^{*r}, U_r) - \mathcal{F}(U_l, U_r^{*l}) \\
&= \mathcal{F}(U_l, U_r) + \partial_1 \mathcal{F}(U_l, U_r)(U_l^{*r} - U_l) - \mathcal{F}(U_l, U_r) - \partial_2 \mathcal{F}(U_l, U_r)(U_r^{*l} - U_r) \\
&\quad + o(|U_l^{*r} - U_l|) + o(|U_r^{*l} - U_r|).
\end{aligned}$$
$$(4.59)$$

But since the conservative numerical flux is consistent, $\mathcal{F}(U, U) = F(U)$ and

$$\partial_1 \mathcal{F}(U, U) + \partial_2 \mathcal{F}(U, U) = F'(U). \qquad (4.60)$$

Performing an expansion in (4.55) gives

$$\begin{aligned}
U_l^{*r} - U_l &= -F'(U_l)^{-1} B(U_l, U_r, \Delta Z)\Delta Z + o(\Delta Z), \\
U_r^{*l} - U_r &= F'(U_r)^{-1} B(U_l, U_r, \Delta Z)\Delta Z + o(\Delta Z),
\end{aligned}$$
$$(4.61)$$

and therefore we obtain, as $\Delta Z, \, U_r - U_l \to 0$,

$$F_r(U_l, U_r, \Delta Z) - F_l(U_l, U_r, \Delta Z) = -B(U_l, U_r, \Delta Z)\Delta Z + o(\Delta Z), \qquad (4.62)$$

which gives the result.   □

The method (4.55)–(4.56) is very simple, but is unfortunately not adapted to treat critical points. When critical points arise, the first difficulty is to solve (4.55). This is possible in the scalar case with a suitable interpretation [14]. Then, another difficulty is that unless the numerical conservative flux $\mathcal{F}(U_l, U_r)$ has a singular jacobian matrix at critical points, the fluxes F_l and F_r have a very low regularity there, which leads to unconsistency. This is definitely bad for systems, for which it is almost impossible to find such numerical fluxes.

Remark 4.6. A variant of the above scheme is

$$
\begin{aligned}
F_l(U_l, U_r, \Delta Z) &= \mathcal{F}(U_l^{*r}, U_r^{*l}) - F(U_l^{*r}) + F(U_l), \\
F_r(U_l, U_r, \Delta Z) &= \mathcal{F}(U_l^{*r}, U_r^{*l}) - F(U_r^{*l}) + F(U_r),
\end{aligned}
\tag{4.63}
$$

where now

$$
\begin{aligned}
F(U_l^{*r}) - F(U_l) + B(U_l, U_r, \Delta Z)(\Delta Z)_+ &= 0, \\
F(U_r) - F(U_r^{*l}) + B(U_l, U_r, \Delta Z)(\Delta Z)_- &= 0,
\end{aligned}
\tag{4.64}
$$

and where $\Delta Z = (\Delta Z)_+ + (\Delta Z)_-$ is some appropriate decomposition. In this case the consistency is merely trivial. The hydrostatic reconstruction scheme explained in Section 4.11 is indeed very close to these formulas.

4.6 Approximate Riemann solvers

The Harten, Lax, Van Leer approximate Riemann solver approach can easily be extended to the system with source (3.3), by considering the variable $\widetilde{U} = (U, Z)$. The relaxation solvers and kinetic solvers exposed in the next sections enter this framework.

The Riemann problem for (3.3) is the problem of solving the system with Riemann data

$$
U^0(x) = \begin{cases} U_l & \text{if } x < 0, \\ U_r & \text{if } x > 0, \end{cases} \qquad Z(x) = \begin{cases} Z_l & \text{if } x < 0, \\ Z_r & \text{if } x > 0. \end{cases}
\tag{4.65}
$$

Since the Z component of the solution is obvious, we shall define only the U component of the approximate Riemann solver.

Definition 4.9. *An approximate Riemann solver for (3.3) is a vector function $R(x/t, U_l, U_r, Z_l, Z_r)$ that is an approximation of the solution to the Riemann problem, in the sense that it must satisfy the basic consistency relation*

$$
R(x/t, U, U, Z, Z) = U,
\tag{4.66}
$$

and the asymptotic conservativity/consistency with the source (4.14), with F_l, F_r defined by

$$
\begin{aligned}
F_l(U_l, U_r, Z_l, Z_r) &= F(U_l, Z_l) - \int_{-\infty}^{0} \Big(R(v, U_l, U_r, Z_l, Z_r) - U_l \Big)\, dv, \\
F_r(U_l, U_r, Z_l, Z_r) &= F(U_r, Z_r) + \int_{0}^{\infty} \Big(R(v, U_l, U_r, Z_l, Z_r) - U_r \Big)\, dv.
\end{aligned}
\tag{4.67}
$$

It is called dissipative with respect to a convex entropy $\widetilde{\eta}$ for (3.3) if

$$
\widetilde{G}_r(U_l, U_r, Z_l, Z_r) - \widetilde{G}_l(U_l, U_r, Z_l, Z_r) \le 0,
\tag{4.68}
$$

where

$$
\begin{aligned}
&\widetilde{G}_l(U_l, U_r, Z_l, Z_r) \\
&= \widetilde{G}(U_l, Z_l) - \int_{-\infty}^{0} \Big(\widetilde{\eta}(R(v, U_l, U_r, Z_l, Z_r), Z_l) - \widetilde{\eta}(U_l, Z_l) \Big)\, dv, \\
&\widetilde{G}_r(U_l, U_r, Z_l, Z_r) \\
&= \widetilde{G}(U_r, Z_r) + \int_{0}^{\infty} \Big(\widetilde{\eta}(R(v, U_l, U_r, Z_l, Z_r), Z_r) - \widetilde{\eta}(U_r, Z_r) \Big)\, dv,
\end{aligned}
\tag{4.69}
$$

and $\widetilde{G}$ is the entropy flux associated to $\widetilde{\eta}$.

With this definition, the numerical scheme (4.1)–(4.2) defined from the fluxes F_l, F_r of an approximate Riemann solver is obviously consistent. The interest of this definition lies in the interpretation of the scheme as the average of an approximate solution, exactly as in the conservative case.

Proposition 4.10. *Consider discrete data (U_i^n, Z_i) at time t_n, and the associated piecewise constant functions $U^n(x)$ and $Z(x)$. Define an approximate solution to (3.1) for $t_n \leq t < t_{n+1}$ by*

$$
U(t, x) = R\left(\frac{x - x_{i+1/2}}{t - t_n}, U_i^n, U_{i+1}^n, Z_i, Z_{i+1} \right) \quad \text{if } x_i < x < x_{i+1}.
\tag{4.70}
$$

This is meaningful under a CFL condition $1/2$, in the sense that

$$
\begin{aligned}
x/t < -\frac{\Delta x_i}{2\Delta t} &\quad \Rightarrow \quad R(x/t, U_i, U_{i+1}, Z_i, Z_{i+1}) = U_i, \\
x/t > \frac{\Delta x_{i+1}}{2\Delta t} &\quad \Rightarrow \quad R(x/t, U_i, U_{i+1}, Z_i, Z_{i+1}) = U_{i+1}.
\end{aligned}
\tag{4.71}
$$

Then the numerical scheme (4.1)–(4.2) is equivalent to the formula

$$
U_i^{n+1} = \frac{1}{\Delta x_i} \int_{x_{i-1/2}}^{x_{i+1/2}} U(t_{n+1} - 0, x)\, dx.
\tag{4.72}
$$

Obviously this convex formula enables to provide invariant domains. This gives also discrete entropy inequalities.

Proposition 4.11. *If an approximate Riemann solver R is entropy dissipative in the sense of (4.68)–(4.69), then the associated numerical scheme is entropy satisfying in the sense of Definition 4.4, with any numerical entropy flux $\widetilde{G}(U_l, U_r, Z_l, Z_r)$ such that*

$$
\widetilde{G}_r(U_l, U_r, Z_l, Z_r) \leq \widetilde{G}(U_l, U_r, Z_l, Z_r) \leq \widetilde{G}_l(U_l, U_r, Z_l, Z_r).
\tag{4.73}
$$

It is also entropy satisfying by interface in the sense of Definition 4.5, for any σ_l, σ_r such that

$$
\begin{aligned}
x/t < \sigma_l &\Rightarrow R(x/t, U_l, U_r, Z_l, Z_r) = U_l, \\
x/t > \sigma_r &\Rightarrow R(x/t, U_l, U_r, Z_l, Z_r) = U_r.
\end{aligned}
\tag{4.74}
$$

4.6.1 Exact solver

Another justification of Definition 4.9 is to prove that the exact Godunov solver satisfies the requirements. In order to do so, we need first to give a precise definition of discrete steady states.

Definition 4.12. *We say that a* (U_l, U_r, Z_l, Z_r) *is an exact discrete steady state if the associated piecewise constant function* $(U(x), Z(x))$ *defined by*

$$U(x) = U_l, \ Z(x) = Z_l, \quad for \ x < 0,$$
$$U(x) = U_r, \ Z(x) = Z_r, \quad for \ x > 0, \tag{4.75}$$

is a generalized solution to (4.4).

We do not make precise the meaning of generalized solution, since a non-conservative product is involved, and resonance can occur. The reader concerned by this question can consult the references proposed in Chapter 3. Indeed in the following the implicit assumptions that we make on this generalized solution can be understood within the (formal) proof. Notice that this is discussed in [28], [43], [29], [3], where exact solvers are used.

Proposition 4.13. *If* $R(x/t, U_l, U_r, Z_l, Z_r)$ *is an exact solver for* (3.1), *in the sense that it gives a generalized solution for all values of the arguments, then it is an approximate solver in the sense of Definition 4.9, and it is well-balanced with respect to exact discrete steady states. Moreover, if the generalized solution satisfies an entropy inequality, then this solver is entropy dissipative in the sense of* (4.68)– (4.69).

Proof. The basic consistency (4.66) is obvious. Let us prove (4.14). Since the second component of the solution is $Z(x)$ as in (4.65), we have

$$Z_x(x) = (Z_r - Z_l)\delta(x), \qquad B(U, Z)Z_x(x) = B^*(Z_r - Z_l)\delta(x), \tag{4.76}$$

for some constant B^*. Thus writing that the component proportional to $\delta(x)$ in (3.1) vanishes, we get

$$F(U_+, Z_r) - F(U_-, Z_l) + B^*(Z_r - Z_l) = 0, \tag{4.77}$$

with

$$U_- = R(0-, U_l, U_r, Z_l, Z_r), \qquad U_+ = R(0+, U_l, U_r, Z_l, Z_r). \tag{4.78}$$

Then, let us integrate the equation over $]0, \Delta t[\times] - \Delta x, 0[$, for some $\Delta t, \Delta x > 0$ satisfying a CFL condition. Since the source does not appear, we get

$$\frac{1}{\Delta x}\int_{-\Delta x}^{0} R(x/\Delta t, U_l, U_r, Z_l, Z_r)\, dx - U_l + \frac{\Delta t}{\Delta x}\Big(F(U_-, Z_l) - F(U_l, Z_l)\Big) = 0. \tag{4.79}$$

Similarly we integrate over $]0, \Delta t[\times]0, \Delta x[$ and get

$$\frac{1}{\Delta x} \int_0^{\Delta x} R(x/\Delta t, U_l, U_r, Z_l, Z_r)\, dx - U_r + \frac{\Delta t}{\Delta x}\Big(F(U_r, Z_r) - F(U_+, Z_r)\Big) = 0.$$
(4.80)

According to the CFL condition, this gives the value of the fluxes defined in (4.67), namely

$$F_l(U_l, U_r, Z_l, Z_r) = F(U_-, Z_l), \qquad F_r(U_l, U_r, Z_l, Z_r) = F(U_+, Z_r). \qquad (4.81)$$

With (4.77), we conclude that (4.14) holds, if B^* is consistent with $B(U)$, which is the least we can ask for a generalized solution.

Next, if the data (U_l, U_r, Z_l, Z_r) are those of a local exact discrete steady state, we have that the exact solution to the Riemann problem is stationary, $R(x/t, U_l, U_r, Z_l, Z_r) = U^0(x)$, and therefore $U_- = U_l$, $U_+ = U_r$, which gives obviously (4.12).

Finally, let us assume that the solution satisfies an entropy inequality

$$\partial_t\, \widetilde{\eta}(U, Z) + \partial_x\, \widetilde{G}(U, Z) \le 0. \qquad (4.82)$$

Taking the component proportional to $\delta(x)$ gives

$$\widetilde{G}(U_+, Z_r) - \widetilde{G}(U_-, Z_l) \le 0. \qquad (4.83)$$

But integrating (4.82) as in (4.79), (4.80), we obtain

$$\widetilde{G}_l(U_l, U_r, Z_l, Z_r) \ge \widetilde{G}(U_-, Z_l), \qquad \widetilde{G}_r(U_l, U_r, Z_l, Z_r) \le \widetilde{G}(U_+, Z_r), \qquad (4.84)$$

and with (4.83) we get (4.68). $\square$

4.6.2 Simple solvers

As in the conservative case, simple solvers are approximate Riemann solvers that have finitely many simple discontinuities. This means that there exists a finite number $m \ge 1$ of speeds

$$\sigma_0 = -\infty < \sigma_1 < \cdots < \sigma_m < \sigma_{m+1} = +\infty, \qquad (4.85)$$

and intermediate states

$$U_0 = U_l, U_1, \ldots, U_{m-1}, U_m = U_r \qquad (4.86)$$

(depending on U_l, U_r, Z_l, Z_r), such that

$$R(x/t, U_l, U_r, Z_l, Z_r) = U_k \quad \text{if} \quad \sigma_k < x/t < \sigma_{k+1}. \qquad (4.87)$$

Since the system (3.3) always has the eigenvalue 0, we must have the speed 0, thus for some m_0,

$$\sigma_{m_0} = 0. \tag{4.88}$$

We complete the intermediate states by

$$\widetilde{U}_k = \begin{cases} (U_k, Z_l) & \text{if } k < m_0, \\ (U_k, Z_r) & \text{if } k \geq m_0. \end{cases} \tag{4.89}$$

The intermediate fluxes F_k, $k = 0, \ldots, m$ are defined by

$$F_k - F_{k-1} = \sigma_k(U_k - U_{k-1}), \ k \neq m_0, \qquad F_0 = F(U_l, Z_l), \ F_m = F(U_r, Z_r). \tag{4.90}$$

The numerical fluxes are obtained as

$$F_l(U_l, U_r, Z_l, Z_r) = F_{m_0-1}, \qquad F_r(U_l, U_r, Z_l, Z_r) = F_{m_0}, \tag{4.91}$$

or equivalently

$$F_l(U_l, U_r, Z_l, Z_r) = F(U_l, Z_l) + \sum_{k=1}^{m_0-1} \sigma_k(U_k - U_{k-1}),$$

$$F_r(U_l, U_r, Z_l, Z_r) = F(U_r, Z_r) - \sum_{k=m_0+1}^{m} \sigma_k(U_k - U_{k-1}). \tag{4.92}$$

The entropy inequality (4.68) becomes

$$\sum_{k=1}^{m} \sigma_k \left(\widetilde{\eta}(\widetilde{U}_k) - \widetilde{\eta}(\widetilde{U}_{k-1}) \right) \geq \widetilde{G}(U_r, Z_r) - \widetilde{G}(U_l, Z_l). \tag{4.93}$$

4.7 Suliciu relaxation solver

We shall not describe here a generalized relaxation framework for solving (3.3), but just write the Suliciu relaxation for Saint Venant system (3.6).

Starting from data $U_l = (\rho_l, \rho_l u_l)$, $U_r = (\rho_r, \rho_r u_r)$, Z_l, Z_r, we solve the relaxation system

$$\begin{cases} \partial_t \rho + \partial_x(\rho u) = 0, \\ \partial_t(\rho u) + \partial_x(\rho u^2 + \pi) + \rho Z_x = 0, \\ \partial_t(\rho \pi/c^2) + \partial_x(\rho \pi u/c^2) + \partial_x u = 0, \\ \partial_t(\rho c) + \partial_x(\rho c u) = 0, \\ \partial_t Z = 0, \end{cases} \tag{4.94}$$

with Riemann initial data, that are completed with

$$\pi_l = p(\rho_l), \quad \pi_r = p(\rho_r), \tag{4.95}$$

and the arbitrary values $c_l, c_r > 0$. The quasilinear system (4.94) has eigenvalues $u - c/\rho$, u (double), $u + c/\rho$, 0, and all are linearly degenerate. Therefore, the meaning of the nonconservative product in (4.94) need not be precised, and it is possible to obtain the solution explicitly. However we shall not describe this in detail here, because this would be too lengthy. Retaining in the solution only the U component, we get a simple approximate Riemann solver for the Saint Venant system, that is entropy dissipative under the subcharacteristic condition (2.119), where c has to be understood as the local value. Integrating the two first equations in (4.94) as in (4.79), (4.80), we get the numerical fluxes

$$F_l = \left(\rho u, \rho u^2 + \pi \right)_-, \qquad F_r = \left(\rho u, \rho u^2 + \pi \right)_+, \qquad (4.96)$$

where $-$ and $+$ denote the values at $x/t = 0-$ or $x/t = 0+$ respectively. A possible choice of c_l and c_r is $c_l/\rho_l = c_r/\rho_r = a > 0$, where a is computed as small as possible satisfying the subcharacteristic conditions. This choice enables to treat the vacuum. The scheme we get is well-balanced with respect to steady states at rest (4.10), because such data give obviously a stationary solution to (4.94) (observe that the third equation becomes stationary when $u = 0$). Therefore, this scheme satisfies all the properties proposed in Section 4.4. We can notice here that the numerical viscosity is minimal, in the sense that when $U_l, U_r \to U$ and $Z_l, Z_r \to Z$, the speeds involved in the approximate Riemann solver tend to the true eigenvalues, since the subcharacteristic condition allows to take $a^2 = p'(\rho)$ in this case.

4.8   Kinetic solver

The kinetic solver of Section 2.5.1 can be extended to treat the source in the Saint Venant system. As proposed in [87] (see also [7] for related methods), we solve

$$\partial_t \underline{f} + \xi \partial_x \underline{f} - Z_x \partial_\xi \underline{f} = 0, \qquad (4.97)$$

with piecewise constant initial maxwellian data, with the maxwellian equilibrium defined by (2.197). Then we define the approximate solution

$$U(t, x) = \int \left(1, \xi \right) \underline{f}(t, x, \xi) \, d\xi, \qquad (4.98)$$

and integrate this over the cell to get the new cell average. This is the nonlocal approach, but of course locally, we have to solve (4.97) with Riemann maxwellian data. One can check that again it gives an approximate Riemann solver for the Saint Venant system with pressure law (2.196), that is entropy dissipative. Integrating (4.97) as in (4.79), (4.80), we get the numerical fluxes

$$F_l = \int \xi \left(1, \xi \right) \underline{f}(x/t = 0-, \xi) \, d\xi, \qquad F_r = \int \xi \left(1, \xi \right) \underline{f}(x/t = 0+, \xi) \, d\xi. \qquad (4.99)$$

The solution $\underline{f}$ is obtained by writing generalized characteristics for (4.97), and this gives the maximal speed in the resolution of the Riemann problem

$$a(U_l, U_r, \Delta Z) = \max\left(\sup_{M(U_l, \xi) \neq 0} \sqrt{\xi^2 + \max(0, -2\Delta Z)}, \right.$$
$$\left. \sup_{M(U_r, \xi) \neq 0} \sqrt{\xi^2 + \max(0, 2\Delta Z)} \right), \qquad (4.100)$$

which is involved in the CFL condition. The interface values are

$$f(x/t = 0-, \xi) = \mathbb{1}_{\xi>0} M(U_l, \xi) + \mathbb{1}_{\xi<0,\, \xi^2-2\Delta Z<0} M(U_l, -\xi)$$
$$+ \mathbb{1}_{\xi<0,\, \xi^2-2\Delta Z>0} M(U_r, -\sqrt{\xi^2 - 2\Delta Z}), \qquad (4.101)$$

$$f(x/t = 0+, \xi) = \mathbb{1}_{\xi<0} M(U_r, \xi) + \mathbb{1}_{\xi>0,\, \xi^2+2\Delta Z<0} M(U_r, -\xi)$$
$$+ \mathbb{1}_{\xi>0,\, \xi^2+2\Delta Z>0} M(U_l, \sqrt{\xi^2 + 2\Delta Z}). \qquad (4.102)$$

The scheme we obtain is well-balanced with respect to steady states at rest, and naturally treats the vacuum. Therefore, this scheme satisfies all the properties proposed in Section 4.4. A practical difficulty that occurs however is that the numerical fluxes cannot be computed explicitly, they involve integrals in ξ that need to be computed numerically.

4.9 VFRoe solver

The VFRoe method for solving the Saint Venant system (3.6) has been introduced in [38]. The method does not enter the framework of the Harten, Lax, Van Leer approximate Riemann solver with source of Section 4.6, but is rather inspired by the formula for the fluxes in the exact solver (4.78), (4.81).

As in the conservative case explained in Section 2.6, we perform a nonlinear change of variables, starting from the quasilinear Saint Venant system. The choice is

$$Y = (\varphi(\rho), u, Z), \qquad (4.103)$$

where $\varphi(\rho)$ is defined by (1.21), and this gives the new quasilinear system

$$\begin{cases} \partial_t \varphi + u\partial_x \varphi + \sqrt{p'(\rho)}\, \partial_x u = 0, \\ \partial_t u + u\partial_x u + \sqrt{p'(\rho)}\, \partial_x \varphi + Z_x = 0, \\ \partial_t Z = 0. \end{cases} \qquad (4.104)$$

Then, we perform the linearization (2.201)–(2.203), which gives

$$\begin{cases} \partial_t \varphi + \widehat{u}\partial_x \varphi + \sqrt{p'(\widehat{\rho})}\, \partial_x u = 0, \\ \partial_t u + \widehat{u}\partial_x u + \sqrt{p'(\widehat{\rho})}\, \partial_x \varphi + Z_x = 0, \\ \partial_t Z = 0, \end{cases} \qquad (4.105)$$

or equivalently

$$\begin{cases} \partial_t(u+\varphi) + (\widehat{u} + \sqrt{p'(\widehat{\rho})})\partial_x(u+\varphi) + Z_x = 0, \\ \partial_t(u-\varphi) + (\widehat{u} - \sqrt{p'(\widehat{\rho})})\partial_x(u-\varphi) + Z_x = 0, \\ \partial_t Z = 0, \end{cases} \qquad (4.106)$$

with Riemann initial data, where

$$\widehat{u} = \frac{u_l + u_r}{2}, \qquad \varphi(\widehat{\rho}) = \frac{\varphi(\rho_l) + \varphi(\rho_r)}{2}. \qquad (4.107)$$

The left and right numerical fluxes are defined as

$$\begin{aligned} F_l &= \left(\frac{(\rho u)_- + (\rho u)_+}{2}, (\rho u^2 + p(\rho))_- \right), \\ F_r &= \left(\frac{(\rho u)_- + (\rho u)_+}{2}, (\rho u^2 + p(\rho))_+ \right), \end{aligned} \qquad (4.108)$$

where $-$ and $+$ denote the values of the solution to (4.105) at $x/t = 0-$ and $x/t = 0+$ respectively, as in (4.78). This formula can easily be proved to be water height conservative, and consistent away from critical points.

The scheme is well-balanced with respect to steady states at rest, because in this case

$$\sqrt{p'(\widehat{\rho})}\left(\varphi(\rho_r) - \varphi(\rho_l)\right) + Z_r - Z_l = 0, \qquad (4.109)$$

leading to a stationary solution for (4.105). Indeed this is true only for a pressure law $p(\rho) = \kappa\rho^\gamma$, but this is enough for applications.

For completeness we give the formula of the Riemann solution to (4.106). Denote the eigenvalues

$$\lambda_1 = \widehat{u} - \sqrt{p'(\widehat{\rho})} < \lambda_2 = \widehat{u} + \sqrt{p'(\widehat{\rho})}. \qquad (4.110)$$

The values at $x/t = 0\pm$ are indexed by $\pm$ as before, and when relevant, an intermediate state for $\lambda_1 < x/t < \lambda_2$ is indexed by '$*$'. Then,
If $0 < \lambda_1 < \lambda_2$,

$$\begin{cases} u^+ = u_l - \dfrac{\Delta Z}{2}(1/\lambda_1 + 1/\lambda_2), \\ \varphi^+ = \varphi_l - \dfrac{\Delta Z}{2}(1/\lambda_2 - 1/\lambda_1), \end{cases} \qquad \begin{cases} u^* = \dfrac{1}{2}(u_l + \varphi_l + u_r - \varphi_r - \Delta Z/\lambda_2), \\ \varphi^* = \dfrac{1}{2}(u_l + \varphi_l - u_r + \varphi_r - \Delta Z/\lambda_2). \end{cases}$$

$$(4.111)$$

If $\lambda_1 < \lambda_2 < 0$,

$$\begin{cases} u^- = u_r + \dfrac{\Delta Z}{2}(1/\lambda_1 + 1/\lambda_2), \\ \varphi^- = \varphi_r + \dfrac{\Delta Z}{2}(1/\lambda_2 - 1/\lambda_1), \end{cases} \qquad \begin{cases} u^* = \dfrac{1}{2}(u_l + \varphi_l + u_r - \varphi_r + \Delta Z/\lambda_1), \\ \varphi^* = \dfrac{1}{2}(u_l + \varphi_l - u_r + \varphi_r - \Delta Z/\lambda_1). \end{cases}$$

$$(4.112)$$

If $\lambda_1 < 0 < \lambda_2$,

$$
\begin{cases}
u^- = \dfrac{1}{2}(u_l + \varphi_l + u_r - \varphi_r + \Delta Z/\lambda_1), \\[2mm]
\varphi^- = \dfrac{1}{2}(u_l + \varphi_l - u_r + \varphi_r - \Delta Z/\lambda_1), \\[2mm]
u^+ = \dfrac{1}{2}(u_l + \varphi_l + u_r - \varphi_r - \Delta Z/\lambda_2), \\[2mm]
\varphi^+ = \dfrac{1}{2}(u_l + \varphi_l - u_r + \varphi_r - \Delta Z/\lambda_2).
\end{cases}
\tag{4.113}
$$

Indeed, in order to get nonnegative densities, one has to put positive parts, as was done in Section 2.6. We observe that when $\Delta Z \equiv Z_r - Z_l \neq 0$, the numerical flux is not only discontinuous when λ_1 or λ_2 crosses 0, but can even take infinite values.

This scheme is extremely simple, and leads to fast execution. However, even if, as was explained in Section 2.6, the variable Y is chosen in order to almost never produce negative densities, this is not always the case. We observe also that non-entropy discontinuities can occur.

4.10 F-wave decomposition method

Generalizations of the Roe method to the case of nonconservative sources are proposed in [12], [13], [100], [25]. The F-wave decomposition method is also a generalization of the Roe method, and has been proposed in [9]. It can be put under the form of a simple approximate Riemann solver, as formulated in Section 4.6.2. Knowing the form (3.4) of the matrix of our system (3.1), and assuming no resonance, one has to take one speed $\sigma_{m_0} = 0$, and the other speeds nonzero. An approximation of the jump in the flux accross the stationary contact discontinuity can be taken as

$$
\delta F_{m_0} = -\widehat{B}(Z_r - Z_l),
\tag{4.114}
$$

for some approximation $\widehat{B}$ of $B(U, Z)$ that needs to be chosen. Then, noticing that for the waves with nonzero eigenvalues, Z has no jump, we can write

$$
\delta F_k = \sigma_k \delta U_k \quad \text{for } k \neq m_0.
\tag{4.115}
$$

Now we write $F(U_r, Z_r) - F(U_l, Z_l) = \sum_{k=1}^{m} \delta F_k$, so that with (4.114) it gives

$$
F(U_r, Z_r) - F(U_l, Z_l) + \widehat{B}(Z_r - Z_l) = \sum_{k \neq m_0} \delta F_k.
\tag{4.116}
$$

It is natural to take δU_k, eigenvector of $\widehat{A}$ for $k \neq m_0$, where $\widehat{A}$ is a diagonalizable matrix (with nonzero eigenvalues) approximating $F_U(U, Z)$, i.e. $\widehat{A}\delta U_k = \sigma_k \delta U_k$. Therefore, by (4.115), δF_k is also an eigenvector,

$$
\widehat{A}\,\delta F_k = \sigma_k \delta F_k \quad \text{for } k \neq m_0.
\tag{4.117}
$$

The relations (4.116)–(4.117) define in a unique way the values σ_k for $k \neq m_0$, which are the distinct eigenvalues of $\widehat{A}$, and the δF_k for $k \neq m_0$. The δU_k are also determined by (4.115), thus we recover the intermediate states by

$$U_k - U_{k-1} = \delta U_k, \quad k \neq m_0. \tag{4.118}$$

Finally, according to (4.92), the fluxes are given by

$$
\begin{aligned}
F_l(U_l, U_r, Z_l, Z_r) &= F(U_l, Z_l) + \sum_{\sigma_k < 0} \delta F_k, \\
F_r(U_l, U_r, Z_l, Z_r) &= F(U_r, Z_r) - \sum_{\sigma_k > 0} \delta F_k.
\end{aligned}
\tag{4.119}
$$

The name of the method, the F-wave decomposition method, comes from the decomposition (4.116) of the jump in fluxes into components δF_k in the eigenspaces of $\widehat{A}$, that directly come into the definition of the numerical fluxes (4.119).

The consistency condition (4.66) is obvious since if $Z_l = Z_r$ and $U_l = U_r$, all δF_k for $k \neq m_0$ vanish by (4.116), thus the δU_k also by (4.115), and $U_k = U_l = U_r$ by (4.118). The consistency with the source (4.14) is also easy to get since by (4.119) and (4.116), $F_r - F_l = F(U_r, Z_r) - F(U_l, Z_l) - \sum_{k \neq m_0} \delta F_k = -\widehat{B}(Z_r - Z_l)$, which reduces therefore to the consistency of $\widehat{B}$.

We notice that in the case without source $Z_l = Z_r$, the method reduces to the Roe method only if $U_{m_0} = U_{m_0-1}$, i.e. $U_l + \sum_{k=1}^{m_0-1} \delta U_k = U_r - \sum_{k=m_0+1}^{m} \delta U_k$, and applying $\widehat{A}$, which is supposed invertible, this gives $\widehat{A}(U_r - U_l) = F(U_r, Z_r) - F(U_l, Z_l)$ i.e. the condition on $\widehat{A}$ to be a Roe matrix. Therefore, this method is naturally a generalization of the Roe method, and a possible choice is $\widehat{A} = A(U_l, U_r, \widehat{Z})$, a Roe matrix obtained by freezing Z to a value $\widehat{Z}$.

The well-balancing property is also easy to obtain from (4.119), it means that $\delta F_k = 0$ for all $k \neq m_0$, or equivalently $F(U_r, Z_r) - F(U_l, Z_l) + \widehat{B}(Z_r - Z_l) = 0$. In the case of the Saint Venant system, one can take

$$\widehat{B} = (0, \widehat{\rho}), \qquad \widehat{\rho} = \frac{p(\rho_r) - p(\rho_l)}{e(\rho_r) + p(\rho_r)/\rho_r - e(\rho_l) - p(\rho_l)/\rho_l}. \tag{4.120}$$

This gives the well-balanced property with respect to the steady states at rest. Moreover the first component of $\widehat{B}$ being 0, this gives the conservativity of the density. However, as in the Roe method, it is not possible to analyze the nonnegativity of density and the entropy inequality.

4.11 Hydrostatic reconstruction scheme

The hydrostatic reconstruction scheme is especially designed to solve Saint Venant type problems. It has been introduced in [6], and is derived from the previous works

[7], [15]. Numerical tests can be found also in [19], [75], [8], [34]. The construction can be seen as a modification of the explicitly well-balanced scheme (4.63), involving the mirror values U_l^{*r}, U_r^{*l} computed by (4.64), or more generally by solving some approximate steaty state equations. The problem in (4.64) is that the occurrence of critical points prevents from finding unique solutions U_l^{*r}, U_r^{*l}. The idea is then to replace the steady state relations (3.11) for the Saint Venant problem by more simple relations,

$$\begin{cases} u = cst, \\ e(\rho) + \dfrac{p(\rho)}{\rho} + Z = cst. \end{cases} \tag{4.121}$$

The interests of these new relations are first that they coincide with the original ones when $u = 0$, the rest steady states, and second that the singularity at critical points is removed when solving the system with arbitrarily given two constants. However, since the relations are not the exact ones, the consistency has to be looked at carefully. Denoting as before $U = (\rho, \rho u)$, the numerical fluxes are defined by

$$\begin{aligned} F_l(U_l, U_r, Z_l, Z_r) &= \mathcal{F}(U_l^*, U_r^*) + \begin{pmatrix} 0 \\ p(\rho_l) - p(\rho_l^*) \end{pmatrix}, \\ F_r(U_l, U_r, Z_l, Z_r) &= \mathcal{F}(U_l^*, U_r^*) + \begin{pmatrix} 0 \\ p(\rho_r) - p(\rho_r^*) \end{pmatrix}, \end{aligned} \tag{4.122}$$

where $\mathcal{F}(U_l, U_r)$ is a given consistent numerical flux for the Saint Venant problem without source. The name "hydrostatic" comes from the fact that in (4.122) and (4.121), only the pressure part of the system is really involved, the advection part being neglected. The reconstructed states U_l^*, U_r^* are defined, according to (4.121), by

$$U_l^* = (\rho_l^*, \rho_l^* u_l), \qquad U_r^* = (\rho_r^*, \rho_r^* u_r), \tag{4.123}$$

$$\begin{aligned} (e + p/\rho)(\rho_l^*) &= ((e + p/\rho)(\rho_l) + Z_l - Z^*)_+, \\ (e + p/\rho)(\rho_r^*) &= ((e + p/\rho)(\rho_r) + Z_r - Z^*)_+, \end{aligned} \tag{4.124}$$

where we recall that $e'(\rho) = p(\rho)/\rho^2$. The positive parts in the right-hand sides of (4.124) are just to ensure that we get some nonnegative densities ρ_l^*, ρ_r^*. The value Z^* is defined by

$$Z^* = \max(Z_l, Z_r), \tag{4.125}$$

and another way of writing (4.124)-(4.125) is, with $\Delta Z = Z_r - Z_l$,

$$\begin{aligned} (e + p/\rho)(\rho_l^*) &= ((e + p/\rho)(\rho_l) - (\Delta Z)_+)_+, \\ (e + p/\rho)(\rho_r^*) &= ((e + p/\rho)(\rho_r) - (-\Delta Z)_+)_+, \end{aligned} \tag{4.126}$$

which shows that F_l, F_r do indeed depend only on the difference ΔZ. The strengh of this scheme is that it is extremely fast, the numerical fluxes being defined directly, without any approximate Riemann solver, and it is indeed also extremely stable.

Proposition 4.14. *Consider a consistent numerical flux $\mathcal{F}$ for the homogeneous Saint Venant problem (i.e. with $Z = cst$), that preserves nonnegativity of the density by interface and satisfies a semi-discrete entropy inequality corresponding to the entropy η in (1.25). Then the scheme defined by the numerical fluxes (4.122), (4.123), (4.126)*

(0) is conservative in density,

(i) preserves the nonnegativity of ρ by interface,

(ii) is well-balanced, i.e. it preserves the discrete steady states at rest (4.10),

(iii) is consistent with the Saint Venant system (3.6),

(iv) satisfies a semi-discrete entropy inequality associated to the entropy $\widetilde{\eta}$ in (3.15).

Proof. Denote the components by $F_l = (F_l^0, F_l^1)$, $F_r = (F_r^0, F_r^1)$. Then the conservativity property (0) is obvious since (4.122) gives that $F_l^0 = F_r^0$.

For property (ii) of well-balancing, consider discrete steady states at rest, i.e. $u_l = u_r = 0$, $(e + p/\rho)(\rho_l) + Z_l = (e + p/\rho)(\rho_r) + Z_r$. Then, (4.124) gives $\rho_l^* = \rho_r^*$, and by (4.123), $U_l^* = U_r^*$. Therefore, by consistency of $\mathcal{F}$ and since $u_l = u_r = 0$, (4.122) gives

$$F_l = F(U_l^*) + \begin{pmatrix} 0 \\ p(\rho_l) - p(\rho_l^*) \end{pmatrix} = F(U_l), \qquad (4.127)$$

$$F_r = F(U_r^*) + \begin{pmatrix} 0 \\ p(\rho_r) - p(\rho_r^*) \end{pmatrix} = F(U_r), \qquad (4.128)$$

which gives (4.12).

To prove the consistency (iii), we have to check the two properties of Definition 4.2. The consistency with the exact flux $F_l(U, U, Z, Z) = F_r(U, U, Z, Z) = F(U)$ is obvious since $U_l^* = U_l$ and $U_r^* = U_r$ whenever $Z_r = Z_l$. For consistency with the source (4.47), we write

$$F_r - F_l = \begin{pmatrix} 0 \\ p(\rho_l^*) - p(\rho_l) + p(\rho_r) - p(\rho_r^*) \end{pmatrix}. \qquad (4.129)$$

Now, we can write $p(\rho_l^*) - p(\rho_l) = [(e + p/\rho)(\rho_l^*) - (e + p/\rho)(\rho_l)]\rho_l^{**}$ for some ρ_l^{**} between ρ_l and ρ_l^*, and $p(\rho_r^*) - p(\rho_r) = [(e + p/\rho)(\rho_r^*) - (e + p/\rho)(\rho_r)]\rho_r^{**}$ for some ρ_r^{**} between ρ_r and ρ_r^*. Then, assuming $\rho > 0$, the positive parts in (4.126) play no role if $\rho_l - \rho$, $\rho_r - \rho$ and ΔZ are small enough. Thus we have $p(\rho_l^*) - p(\rho_l) = -\rho_l^{**}(\Delta Z)_+$, $p(\rho_r^*) - p(\rho_r) = -\rho_r^{**}(-\Delta Z)_+$, which gives (4.47). In the special case $\rho = 0$, the positive parts in (4.126) can play a role only when $(e + p/\rho)(\rho_l) = O(\Delta Z)$, or respectively $(e + p/\rho)(\rho_r) = O(\Delta Z)$, and we conclude that (4.47) always holds, proving (iii).

Let us now prove the nonnegativity statement (i). It is a consequence of the property

$$\rho_l^* \leq \rho_l, \qquad \rho_r^* \leq \rho_r, \qquad (4.130)$$

that comes directly from the definition (4.126). By assumption the numerical flux $\mathcal{F}$ preserves the nonnegativity or ρ by interface. By Definition 2.4, this means that there exists some $\sigma_l(U_l, U_r) < 0 < \sigma_r(U_l, U_r)$ such that

$$\rho_l + \frac{\mathcal{F}^0(U_l, U_r) - \rho_l u_l}{\sigma_l(U_l, U_r)} \geq 0, \qquad \rho_r + \frac{\mathcal{F}^0(U_l, U_r) - \rho_r u_r}{\sigma_r(U_l, U_r)} \geq 0, \qquad (4.131)$$

for any U_l and U_r (with nonnegative densities ρ_l, ρ_r). This implies in particular that

$$\rho_l^* + \frac{\mathcal{F}^0(U_l^*, U_r^*) - \rho_l^* u_l}{\sigma_l(U_l^*, U_r^*)} \geq 0, \qquad \rho_r^* + \frac{\mathcal{F}^0(U_l^*, U_r^*) - \rho_r^* u_r}{\sigma_r(U_l^*, U_r^*)} \geq 0. \qquad (4.132)$$

But since necessarily $1 - u_l/\sigma_l(U_l^*, U_r^*) \geq 0$, $1 - u_r/\sigma_r(U_l^*, U_r^*) \geq 0$, we deduce with (4.130) that

$$\rho_l + \frac{F_l^0(U_l, U_r, Z_l, Z_r) - \rho_l u_l}{\sigma_l(U_l^*, U_r^*)} \geq 0, \qquad \rho_r + \frac{F_r^0(U_l, U_r, Z_l, Z_r) - \rho_r u_r}{\sigma_r(U_l^*, U_r^*)} \geq 0,$$

$$(4.133)$$

which means that the scheme preserves the nonnegativity of ρ by interface. Note in particular that the speeds are $\sigma_l(U_l^*, U_r^*)$ and $\sigma_r(U_l^*, U_r^*)$, thus the CFL condition associated to the scheme is the one of $\mathcal{F}$ corresponding to the data U_l^* and U_r^*. Surprisingly, since the sound speed is usually a monotone function of the density and because of (4.130), this is slightly less restrictive than expected.

Let us finally prove the entropy inequality statement (iv). Here "semi-discrete" refers to the limit when the timestep Δt tends to 0, and therefore we have to prove the existence of some consistent numerical entropy flux $\widetilde{G}(U_l, U_r, Z_l, Z_r)$ such that the inequalities (4.40), (4.41) hold. By assumption, $\mathcal{F}$ satisfies a semi-discrete entropy inequality for the problem without source, thus there exists some numerical entropy flux $\mathcal{G}$, consistent with G in (1.27), such that according to (2.25), (2.26)

$$G(U_r) + \eta'(U_r)(\mathcal{F}(U_l, U_r) - F(U_r)) \leq \mathcal{G}(U_l, U_r),$$
$$\mathcal{G}(U_l, U_r) \leq G(U_l) + \eta'(U_l)(\mathcal{F}(U_l, U_r) - F(U_l)). \qquad (4.134)$$

Note that since differentiation is with respect to $U = (\rho, \rho u)$,

$$\eta'(U) = \left(e(\rho) + p(\rho)/\rho - u^2/2, u \right). \qquad (4.135)$$

Let us start by applying (4.134) to U_l^*, U_r^*. We get

$$G(U_r^*) + \eta'(U_r^*)(\mathcal{F}(U_l^*, U_r^*) - F(U_r^*)) \leq \mathcal{G}(U_l^*, U_r^*),$$
$$\mathcal{G}(U_l^*, U_r^*) \leq G(U_l^*) + \eta'(U_l^*)(\mathcal{F}(U_l^*, U_r^*) - F(U_l^*)). \qquad (4.136)$$

We are going to prove that (4.40), (4.41) hold with

$$\widetilde{G}(U_l, U_r, Z_l, Z_r) = \mathcal{G}(U_l^*, U_r^*) + \mathcal{F}^0(U_l^*, U_r^*)Z^*. \qquad (4.137)$$

This formula is obviously consistent with $\widetilde{G}$, recall that $\widetilde{G} = G + \rho u Z$ and $\widetilde{\eta} = \eta + \rho Z$ from (3.15). Since both inequalities are obtained exactly in the same way, let us prove only the left inequality (4.41). By comparison with (4.136), it is enough to prove that

$$
\begin{aligned}
&G(U_l^*) + \eta'(U_l^*)\big(\mathcal{F}(U_l^*, U_r^*) - F(U_l^*)\big) + \mathcal{F}^0(U_l^*, U_r^*)Z^* \\
&\leq G(U_l) + \eta'(U_l)(F_l - F(U_l)) + \mathcal{F}^0(U_l^*, U_r^*)Z_l.
\end{aligned}
\tag{4.138}
$$

Adopting the shorthand notation $\mathcal{F}(U_l^*, U_r^*) = (\mathcal{F}^0, \mathcal{F}^1)$, this inequality can be written

$$
\begin{aligned}
&(u_l^2/2 + (e + p/\rho)(\rho_l^*))\rho_l^* u_l + ((e + p/\rho)(\rho_l^*) - u_l^2/2)(\mathcal{F}^0 - \rho_l^* u_l) \\
&\quad + u_l(\mathcal{F}^1 - \rho_l^* u_l^2 - p(\rho_l^*)) + \mathcal{F}^0(Z^* - Z_l) \\
&\leq (u_l^2/2 + (e + p/\rho)(\rho_l))\rho_l u_l + ((e + p/\rho)(\rho_l) - u_l^2/2)(\mathcal{F}^0 - \rho_l u_l) \\
&\quad + u_l(F_l^1 - \rho_l u_l^2 - p(\rho_l)),
\end{aligned}
\tag{4.139}
$$

or after simplification

$$
u_l(\mathcal{F}^1 - p(\rho_l^*)) + \mathcal{F}^0((e+p/\rho)(\rho_l^*) - (e+p/\rho)(\rho_l) + Z^* - Z_l) \leq u_l(F_l^1 - p(\rho_l)). \tag{4.140}
$$

Since $F_l^1 - p(\rho_l) = \mathcal{F}^1 - p(\rho_l^*)$ by definition of F_l in (4.122), our inequality finally reduces to

$$
\mathcal{F}^0(U_l^*, U_r^*)((e + p/\rho)(\rho_l^*) - (e + p/\rho)(\rho_l) + Z^* - Z_l) \leq 0. \tag{4.141}
$$

Now, according to (4.124), when this quantity is nonzero, we have $\rho_l^* = 0$ and the expression between parentheses is nonnegative. But since $\mathcal{F}$ preserves nonnegativity of density, we have by (4.132) that $\mathcal{F}^0(\rho_l^* = 0, u_l, \rho_r^*, u_r) \leq 0$ and we conclude that (4.141) always holds. This completes the proof of (iv). $\qquad\square$

The hydrostatic reconstruction scheme satisfies all the properties of Section 4.4, except that the entropy inequality is only semi-discrete. One can check that indeed it does not satisfy a fully discrete entropy inequality. There exist some data with $(e + p/\rho)(\rho_i) + Z_i = cst$, $u_i = cst \neq 0$ such that for any $\Delta t > 0$, the fully discrete entropy inequality $\widetilde{\eta}(U_i^{n+1}, Z_i) - \widetilde{\eta}(U_i^n, Z_i) + \frac{\Delta t}{\Delta x_i}(\widetilde{G}_{i+1/2} - \widetilde{G}_{i-1/2}) \leq 0$ is violated. However these data are not preserved by the scheme. The consequence is that in practice we do not observe instabilities, as long as ρ_i remains nonnegative, which is the case under the CFL condition of the homogeneous numerical flux $\mathcal{F}$.

Additionally to its low cost, the hydrostatic reconstruction scheme has the property to be easily adaptable to systems of the same type as the Saint Venant problem. We give two examples of such extensions: the Saint Venant problem with variable pressure, and the nozzle problem. We mention also that the method can be modified to preserve exactly all the subsonic steady states, this is explained in [21].

4.11.1 Saint Venant problem with variable pressure

This problem occurs in the same physical situation as the Saint Venant problem, when the topography slope is not supposed small, see [20]. It reads as

$$\begin{cases} \partial_t \rho + \partial_x(\rho u) = 0, \\ \partial_t(\rho u) + \partial_x(\rho u^2 + p(\rho)\iota) + \rho g z_x + \rho e(\rho)\iota_x = 0, \end{cases} \tag{4.142}$$

where $g > 0$, $z = z(x)$, and $\iota = \iota(x) > 0$. As before the pressure $p(\rho)$ is increasing, and $e'(\rho) = p(\rho)/\rho^2$. This system takes the form (3.1), with $U = (\rho, \rho u)$ and $Z = (gz, \iota)$ which is here two-dimensional. The interesting feature is that here the conservative part of the equation also depends on Z. Combining the two equations in (4.142), we get for smooth solutions

$$\partial_t u + \partial_x \left(u^2/2 + (e(\rho) + p(\rho)/\rho)\iota + gz \right) = 0, \tag{4.143}$$

and thus multiplying (4.143) by ρu and the first line of (4.142) by $u^2/2 + (e(\rho) + p(\rho)/\rho)\iota + gz$, it yields by addition the energy conservation, that becomes an inequality for weak solutions

$$\partial_t \left(\rho u^2/2 + \rho e(\rho)\iota + \rho gz \right) + \partial_x \left[\left(\rho u^2/2 + (\rho e(\rho) + p(\rho))\iota + \rho gz \right) u \right] \leq 0. \tag{4.144}$$

This shows that $\tilde{\eta} = \rho u^2/2 + \rho e(\rho)\iota + \rho gz$ is an entropy, with entropy flux $\tilde{G} = (\rho u^2/2 + (\rho e(\rho) + p(\rho))\iota + \rho gz)u$. From the first line of (4.142) and (4.143) we obtain the steady states, characterized by

$$\begin{cases} \rho u = cst, \\ \dfrac{u^2}{2} + \left(e(\rho) + \dfrac{p(\rho)}{\rho} \right)\iota + gz = cst. \end{cases} \tag{4.145}$$

In particular, we are again especially interested in steady states at rest where

$$\begin{cases} u = 0, \\ \left(e(\rho) + \dfrac{p(\rho)}{\rho} \right)\iota + gz = cst. \end{cases} \tag{4.146}$$

In order to apply the hydrostatic reconstruction method to this problem, we consider a numerical flux for the problem *without source*. Since here the flux function depends on Z via ι, $F(U, \iota) = (\rho u, \rho u^2 + p(\rho)\iota)$, we have to give a precise sense to this, and this means that we solve the problem (4.142) when $Z = cst$. Thus $\iota > 0$ is just a parameter, and a numerical flux for this problem is a function $\mathcal{F}(U_l, U_r, \iota)$. Then the hydrostatic reconstruction scheme for the full problem with source is obtained with the left and right numerical fluxes computed as follows,

$$F_l(U_l, U_r, Z_l, Z_r) = \mathcal{F}(U_l^*, U_r^*, \iota^*) + \begin{pmatrix} 0 \\ p(\rho_l)\iota_l - p(\rho_l^*)\iota^* \end{pmatrix},$$

$$F_r(U_l, U_r, Z_l, Z_r) = \mathcal{F}(U_l^*, U_r^*, \iota^*) + \begin{pmatrix} 0 \\ p(\rho_r)\iota_r - p(\rho_r^*)\iota^* \end{pmatrix}, \tag{4.147}$$

where

$$U_l^* = (\rho_l^*, \rho_l^* u_l), \qquad U_r^* = (\rho_r^*, \rho_r^* u_r), \tag{4.148}$$

$$(e + p/\rho)(\rho_l^*) = \Big((e + p/\rho)(\rho_l)\iota_l + g z_l - g z^* \Big)_+ / \iota^*,$$
$$(e + p/\rho)(\rho_r^*) = \Big((e + p/\rho)(\rho_r)\iota_r + g z_r - g z^* \Big)_+ / \iota^*, \tag{4.149}$$

$$z^* = \max(z_l, z_r), \qquad \iota^* = \max(\iota_l, \iota_r). \tag{4.150}$$

Proposition 4.15. *Consider a consistent numerical flux $\mathcal{F}(U_l, U_r, \iota)$ for the problem (4.142) without source, i.e. with $z = cst$ and $\iota = cst$, that preserves nonnegativity of the density by interface and satisfies a semi-discrete entropy inequality corresponding to the entropy $\eta = \rho u^2/2 + \rho e(\rho)\iota$. Then the scheme defined by the numerical fluxes (4.147)–(4.150)*
(0) is conservative in density,
(i) preserves the nonnegativity of ρ by interface,
(ii) is well-balanced, i.e. it preserves the discrete steady states at rest,
(iii) is consistent with the Saint Venant system with variable pressure (4.142),
(iv) satisfies a semi-discrete entropy inequality associated to the entropy $\widetilde{\eta}$ in (4.144).

We do not give the proof of this proposition, which follows exactly the one of Proposition 4.14. We just mention that the choice (4.150) is made to have

$$\rho_l^* \le \rho_l, \qquad \rho_r^* \le \rho_r, \tag{4.151}$$

which ensures as before the nonnegativity property. For the semi-discrete entropy inequality, the numerical entropy flux is obtained as $\widetilde{G}(U_l, U_r, Z_l, Z_r) = \mathcal{G}(U_l^*, U_r^*, \iota^*) + \mathcal{F}^0(U_l^*, U_r^*, \iota^*) g z^*$, where $\mathcal{G}$ is the numerical entropy flux for the problem without source.

4.11.2   Nozzle problem

This system describes the evolution of a gas in a nozzle, see [73], [4], and writes as

$$\begin{cases} \partial_t(Z\rho) + \partial_x(Z\rho u) = 0, \\ \partial_t(Z\rho u) + \partial_x\big(Z(\rho u^2 + p(\rho))\big) = p(\rho) Z_x, \end{cases} \tag{4.152}$$

where $Z = Z(x) > 0$ is the nozzle section, and as usual $p'(\rho) > 0$. The system is of the form (3.1) with conservative variables $Z\rho$ and $Z\rho u$. For smooth solutions, a combination of the two equations gives

$$\partial_t u + \partial_x(u^2/2 + e(\rho) + p(\rho)/\rho) = 0, \tag{4.153}$$

with as before $e'(\rho) = p(\rho)/\rho^2$. Then, multiplying (4.153) by $Z\rho u$ and the first line of (4.152) by $u^2/2 + e(\rho) + p(\rho)/\rho$, the sum gives the energy inequality

$$\partial_t\Big(Z(\rho u^2/2 + \rho e(\rho)) \Big) + \partial_x\Big(Z(\rho u^2/2 + \rho e(\rho) + p(\rho))u \Big) \le 0. \tag{4.154}$$

With (4.153) and the first line of (4.152), the steady states are given by

$$
\begin{cases}
Z\rho u = cst, \\
\dfrac{u^2}{2} + e(\rho) + \dfrac{p(\rho)}{\rho} = cst.
\end{cases}
\tag{4.155}
$$

The steady states at rest are very simple for this system,

$$
\begin{cases}
u = 0, \\
\rho = cst.
\end{cases}
\tag{4.156}
$$

In order to apply the hydrostatic reconstruction method to this problem, we first consider the problem without source, i.e. when $Z = cst$. In this case, (4.152) simplifies to the usual isentropic gas dynamics system. Therefore, let us consider given a numerical flux $\mathcal{F}(U_l, U_r)$ for the isentropic gas dynamics system, the corresponding numerical flux for (4.152) with constant Z being just $\mathcal{F}(U_l, U_r, Z) = Z\mathcal{F}(U_l, U_r)$. Then, for the full problem (4.152), we denote $U = (\rho, \rho u)$ (which is not the conservative variable here), and we define the numerical fluxes by

$$
F_l(U_l, U_r, Z_l, Z_r) = Z^*\mathcal{F}(U_l, U_r) + \begin{pmatrix} 0 \\ Z_l p(\rho_l) - Z^* p(\rho_l) \end{pmatrix},
$$
$$
F_r(U_l, U_r, Z_l, Z_r) = Z^*\mathcal{F}(U_l, U_r) + \begin{pmatrix} 0 \\ Z_r p(\rho_r) - Z^* p(\rho_r) \end{pmatrix},
\tag{4.157}
$$

$$
Z^* = \min(Z_l, Z_r).
\tag{4.158}
$$

Indeed, this means that we have chosen $U_l^* = U_l$ and $U_r^* = U_r$, which is coherent with (4.156). The choice (4.158) is motivated as before by the nonnegativity property.

Proposition 4.16. *Consider a consistent numerical flux $\mathcal{F}$ for the isentropic gas dynamics system, that preserves nonnegativity of the density by interface and satisfies a semi-discrete entropy inequality corresponding to the entropy η in (1.25). Then the scheme defined by the numerical fluxes (4.157), (4.158)*
(0) is conservative in density,
(i) preserves the nonnegativity of ρ by interface,
(ii) is well-balanced, i.e. it preserves the discrete steady states at rest (4.156),
(iii) is consistent with the system (4.152),
(iv) satisfies a semi-discrete entropy inequality associated to the entropy $\widetilde{\eta} = Z\eta$.

Proof. Only (i) and (iv) are nontrivial. Let us prove (i). Since the numerical flux $\mathcal{F}$ is nonnegative by interface, according to Definition 2.4 there exists some $\sigma_l < 0 < \sigma_r$ such that

$$
\rho_l + \frac{\mathcal{F}^0(U_l, U_r) - \rho_l u_l}{\sigma_l} \geq 0, \qquad \rho_r + \frac{\mathcal{F}^0(U_l, U_r) - \rho_r u_r}{\sigma_r} \geq 0.
\tag{4.159}
$$

Then, since $1 - u_l/\sigma_l \geq 0$ and $1 - u_r/\sigma_r \geq 0$, multiplying by Z^* and using that $Z^* \leq Z_l$ and $Z^* \leq Z_r$ we get

$$Z_l \rho_l + \frac{Z^* \mathcal{F}^0(U_l, U_r) - Z_l \rho_l u_l}{\sigma_l} \geq 0, \qquad Z_r \rho_r + \frac{Z^* \mathcal{F}^0(U_l, U_r) - Z_r \rho_r u_r}{\sigma_r} \geq 0, \tag{4.160}$$

which is the nonnegativity property for our scheme.

Next, let us prove (iv). We have $F(U, Z) = ZF(U)$, $\widetilde{\eta}(U, Z) = Z\eta(U)$, $\widetilde{G}(U, Z) = ZG(U)$, and $\widetilde{\eta}'(U, Z) = \eta'(U) = (e + p/\rho - u^2/2, u)$ because the differentiation is with respect to the conservative variable ZU. By assumption we have a numerical entropy flux $\mathcal{G}(U_l, U_r)$ such that

$$\begin{aligned} G(U_r) + \eta'(U_r)(\mathcal{F}(U_l, U_r) - F(U_r)) &\leq \mathcal{G}(U_l, U_r), \\ \mathcal{G}(U_l, U_r) &\leq G(U_l) + \eta'(U_l)(\mathcal{F}(U_l, U_r) - F(U_l)). \end{aligned} \tag{4.161}$$

Multiplying (4.161) by Z^* and using the identity

$$G(U) + \eta'(U) \left(\begin{pmatrix} 0 \\ p(\rho) \end{pmatrix} - F(U) \right) = 0, \tag{4.162}$$

we get

$$\begin{aligned} Z_r G(U_r) + \eta'(U_r)(F_r - Z_r F(U_r)) &\leq Z^* \mathcal{G}(U_l, U_r), \\ Z^* \mathcal{G}(U_l, U_r) &\leq Z_l G(U_l) + \eta'(U_l)(F_l - Z_l F(U_l)), \end{aligned} \tag{4.163}$$

which means that the scheme satisfies a semi-discrete entropy intequality, with numerical entropy flux $\widetilde{G}(U_l, U_r, Z_l, Z_r) = Z^* \mathcal{G}(U_l, U_r)$.   □

4.12 Additional source terms

The method that we propose here to treat additional zero-order source terms with the well-balanced property is general, but for clarity we restrict here to the Saint Venant problem. A more general formulation is proposed in Chapter 5 in the multidimensional context. This method has been showed to be very efficient when dealing with Coriolis force in [19], [75].

Consider the Saint Venant system with topography and external force f,

$$\begin{cases} \partial_t \rho + \partial_x(\rho u) = 0, \\ \partial_t(\rho u) + \partial_x(\rho u^2 + p(\rho)) + \rho Z_x = \rho f, \end{cases} \tag{4.164}$$

where $Z = Z(x)$, $f = f(t, x)$. One can think in particular of a nonlinear coupling like $f(t, x) = g(\rho, u)$. Now the velocity equation writes

$$\partial_t u + \partial_x(u^2/2 + e(\rho) + p(\rho)/\rho + Z) = f, \tag{4.165}$$

and the steady states at rest are given by

$$u = 0, \qquad \partial_x(e(\rho) + p(\rho)/\rho + Z) = f. \tag{4.166}$$

The idea to solve (4.164) is to identify the system as the usual Saint Venant problem with a new topography $Z + B$, where $B_x = -f$. Now, B depends also on time while it should be time independent, but when using discrete times t_n, we can freeze the value on a time interval, thus we take $B_x^n = -f^n$ and solve the Saint Venant system on the time interval (t_n, t_{n+1}) with topography $Z + B^n$. Note that for stationary solutions, this approximation is exact, thus the stationary solutions are preserved by this procedure.

At the fully discrete level, this is done as follows. We define

$$\Delta B_{i+1/2}^n = -f_{i+1/2}^n \Delta x_{i+1/2}, \tag{4.167}$$

where $\Delta x_{i+1/2} = x_{i+1} - x_i$, and update $U = (\rho, \rho u)$ via

$$U_i^{n+1} - U_i^n + \frac{\Delta t}{\Delta x_i}(F_{i+1/2-} - F_{i-1/2+}) = 0, \tag{4.168}$$

with

$$\begin{aligned}
F_{i+1/2-} &= F_l(U_i, U_{i+1}, \Delta Z_{i+1/2} + \Delta B_{i+1/2}^n), \\
F_{i+1/2+} &= F_r(U_i, U_{i+1}, \Delta Z_{i+1/2} + \Delta B_{i+1/2}^n),
\end{aligned} \tag{4.169}$$

where $\Delta Z_{i+1/2} = Z_{i+1} - Z_i$ and the numerical fluxes $F_l(U_l, U_r, \Delta Z)$, $F_r(U_l, U_r, \Delta Z)$ are associated to the usual Saint Venant problem. If the numerical fluxes F_l, F_r are consistent with the Saint Venant system, then the new scheme (4.167)–(4.169) is consistent with (4.164), in a sense that is an obvious generalization of Definition 4.2. Moreover, if the numerical fluxes F_l, F_r are well-balanced with respect to steady states at rest, then the new scheme (4.167)–(4.169) automatically preserves the data satisfying

$$u_i = 0, \quad e(\rho_{i+1}) + p(\rho_{i+1})/\rho_{i+1} + Z_{i+1} = e(\rho_i) + p(\rho_i)/\rho_i + Z_i + f_{i+1/2}\Delta x_{i+1/2}, \tag{4.170}$$

which can be considered as discrete steady states approximating the relations (4.166). Concerning discrete entropy inequalities, there is no reason in general to have a special compatibility with this method, but however if f is bounded, this should not be a problem because the right-hand side is a lower-order term that should not influence the global stability.

4.12.1 Saint Venant problem with Coulomb friction

In order to illustrate the treatment of additional source terms, let us consider the Saint Venant system with Coulomb friction

$$\begin{cases} \partial_t \rho + \partial_x(\rho u) = 0, \\ \partial_t(\rho u) + \partial_x(\rho u^2 + p(\rho)) + \rho Z_x = -\rho g \mu \operatorname{sgn} u, \end{cases} \tag{4.171}$$

where $\mu \geq 0$ is the friction coefficient. Such a model arises in the modeling of debris avalanches, see for example [22]. The term $\operatorname{sgn} u$ has to be understood as multivalued: sgn 0 can be any value in $[-1, 1]$. A rigorous definition of the meaning of (4.171) is indeed that there must exist some function $f(t, x)$ such that

$$\begin{cases} \partial_t \rho + \partial_x(\rho u) = 0, \\ \partial_t(\rho u) + \partial_x(\rho u^2 + p(\rho)) + \rho Z_x = \rho f, \end{cases} \tag{4.172}$$

and

$$\begin{cases} |f(t, x)| \leq g\mu, \\ u(t, x) \neq 0 \;\Rightarrow\; f(t, x) = -g\mu \operatorname{sgn} u(t, x). \end{cases} \tag{4.173}$$

An existence result within this formulation can be found in [55]. By putting together (4.166) and (4.173) we obtain he steady states at rest, given by

$$u = 0, \qquad \left| \partial_x(e(\rho) + p(\rho)/\rho + Z) \right| \leq g\mu. \tag{4.174}$$

In order to apply the scheme (4.167)-(4.169), we just need to define a consistent value for $f_{i+1/2}^n$. This is done by setting

$$f_{i+1/2}^n = -\operatorname*{proj}_{g\mu} \left(\frac{e(\rho_i) + p(\rho_i)/\rho_i - e(\rho_{i+1}) - p(\rho_{i+1})/\rho_{i+1} - \Delta Z_{i+1/2}}{\Delta x_{i+1/2}} + \frac{u_{i+1/2}}{\Delta t} \right), \tag{4.175}$$

where

$$\operatorname*{proj}_{g\mu}(X) = \begin{cases} X & \text{if } |X| \leq g\mu, \\ g\mu \dfrac{X}{|X|} & \text{if } |X| > g\mu, \end{cases} \tag{4.176}$$

and for example

$$u_{i+1/2} = \frac{\rho_i u_i + \rho_{i+1} u_{i+1}}{\rho_i + \rho_{i+1}}. \tag{4.177}$$

These formula define a value of $f_{i+1/2}^n$ which is consistent with (4.173) because $|f_{i+1/2}^n| \leq g\mu$, and if $u_{i+1/2} \neq 0$, then for Δt small enough the ratio $u_{i+1/2}/\Delta t$ in (4.175) will dominate the other term, giving $f_{i+1/2}^n = -g\mu \operatorname{sgn} u_{i+1/2}$. This construction is also well-balanced. Indeed the data such that

$$u_i = 0, \qquad \left| e(\rho_i) + p(\rho_i)/\rho_i - e(\rho_{i+1}) - p(\rho_{i+1})/\rho_{i+1} - \Delta Z_{i+1/2} \right| \leq g\mu \Delta x_{i+1/2} \tag{4.178}$$

are preserved by the scheme, because this yields $\Delta Z_{i+1/2} + \Delta B_{i+1/2}^n = e(\rho_i) + p(\rho_i)/\rho_i - e(\rho_{i+1}) - p(\rho_{i+1})/\rho_{i+1}$, which is the relation that gives no evolution in (4.168)–(4.169).

Another interesting property of (4.175) can be seen if we consider particular solutions of (4.171) that satisfy $\rho = cst$, $Z_x = cst$, and $u = u(t)$. Then the system simplifies to

$$\frac{du}{dt} + Z_x = -g\mu \operatorname{sgn} u. \tag{4.179}$$

This ordinary differential equation with multivalued right-hand side has indeed unique solutions because the right-hand side is a monotone operator. The solutions have discontinuous derivative, thus the usual forward Euler scheme suffers from very low accuracy when the derivative is discontinuous. On the contrary, our scheme is here

$$u^{n+1} = u^n + \Delta t(f^n - Z_x), \qquad f^n = -\operatorname*{proj}_{g\mu}\left(-Z_x + u^n/\Delta t\right). \qquad (4.180)$$

It has the property that when $|Z_x| < g\mu$ and u^n is small enough, it gives directly the exact solution $u^{n+1} = 0$, and then it remains at 0. However, it is still imprecise in the case $|Z_x| > g\mu$ where the solution has to cross 0 with a discontinuity in the derivative. In order to have an exact resolution in both cases, one have to use a variant of the above scheme, replacing (4.175) by

$$f^n_{i+1/2} = -\varphi_{g\mu}\left(\frac{e(\rho_i) + p(\rho_i)/\rho_i - e(\rho_{i+1}) - p(\rho_{i+1})/\rho_{i+1} - \Delta Z_{i+1/2}}{\Delta x_{i+1/2}}, \frac{u_{i+1/2}}{\Delta t}\right),$$
$$(4.181)$$

where

$$\varphi_{g\mu}(X, Y) = \operatorname*{proj}_{g\mu}\left(\operatorname*{proj}_{g\mu}(X) + \frac{2}{1 + \max(1, -X \cdot Y/g\mu|Y|)}Y\right), \qquad (4.182)$$

instead of $\varphi_{g\mu}(X, Y) = \operatorname*{proj}_{g\mu}(X + Y)$ before.

4.13 Second-order extension

Few authors have proposed second-order accurate well-balanced schemes. We shall give here a general method derived from [66], [67], which is a natural extension of Section 2.8. It shares also ideas with [76]. The method uses a second-order reconstruction operator, and consistent first-order fluxes F_l, F_r in the sense of Definition 4.2. Since second-order accuracy in time will be obtained as usual by the Heun method (2.261), it is enough to build a scheme that is second-order in space only.

The second-order reconstruction operator is as in Definition 2.25, but acts on the sequence $(U_i, Z_i)_{i \in \mathbb{Z}}$, thus giving values $(U_{i+1/2-}, Z_{i+1/2-})$ and $(U_{i+1/2+}, Z_{i+1/2+})$. We only modify slightly the conservativity (2.230) that is assumed only for U (not for Z).
The second-order scheme is defined by

$$U_i^{n+1} - U_i^n + \frac{\Delta t}{\Delta x_i}(F_{i+1/2-} - F_{i-1/2+} - \delta F_i) = 0, \qquad (4.183)$$

with

$$F_{i+1/2-} = F_l\left(U^n_{i+1/2-}, U^n_{i+1/2+}, Z^n_{i+1/2-}, Z^n_{i+1/2+}\right),$$
$$F_{i+1/2+} = F_r\left(U^n_{i+1/2-}, U^n_{i+1/2+}, Z^n_{i+1/2-}, Z^n_{i+1/2+}\right), \qquad (4.184)$$
$$\delta F_i = F_c\left(U^n_{i-1/2+}, U^n_{i+1/2-}, Z^n_{i-1/2+}, Z^n_{i+1/2-}\right),$$

and the function F_c needs to be chosen. We remark that although Z_i does not change with time, the interface values $Z^n_{i+1/2\pm}$ can be time dependent if the reconstruction operator is not performed componentwise. We define

$$\Delta Z^n_{i+1/2} = Z^n_{i+1/2+} - Z^n_{i+1/2-}, \qquad \Delta Z^n_i = Z^n_{i+1/2-} - Z^n_{i-1/2+}. \qquad (4.185)$$

4.13.1 Second-order accuracy

The idea in (4.183)–(4.184) is that depending on the smoothness of the computed solution, the source term is either discretized at the interfaces or at the centers of the cells. One one hand, if the solution has large discontinuities, the reconstruction step should give $(U, Z)^n_{i+1/2-} = (U, Z)^n_{i-1/2+} = (U^n_i, Z_i)$, thus reducing to the first-order scheme. This indicates that we have to impose

$$F_c(U, U, Z, Z) = 0. \qquad (4.186)$$

On the other hand, if the solution is smooth, the interface jumps $(U, Z)_{i+1/2+} - (U, Z)_{i+1/2-}$ should be small, and the source cannot appear at the interface, because (4.14) gives nothing if $\Delta Z^n_{i+1/2}$ is small. It has to be handled by the centered term δF_i. In order that this term gives a second-order resolution, we shall require that

$$\begin{aligned}
&F_c(U_l, U_r, Z_l, Z_r) \\
&= -\left(B\left(\frac{U_l + U_r}{2}, \frac{Z_l + Z_r}{2} \right) + O(|U_r - U_l|^2 + |Z_r - Z_l|^2) \right)(Z_r - Z_l),
\end{aligned} \qquad (4.187)$$

as $U_r - U_l \to 0$ and $Z_r - Z_l \to 0$. This condition implies that $F_c(U_l, U_r, Z, Z) = 0$, which is stronger than (4.186).

In view of (4.14), a candidate for F_c would be $F_c = F_r - F_l$. But in general there is no chance that it satisfies the centered expansion (4.187), because $F_r - F_l$ involves somehow some upwinding. This choice would be also good for invariant domains.

Lemma 4.17. *If under a CFL condition the numerical fluxes preserve a convex invariant domain $U \in \mathcal{U}$, and if the reconstruction also preserves this invariant domain, then under the half original CFL condition, the second-order scheme (4.183)–(4.184) with $F_c = F_r - F_l$ also preserves this invariant domain.*

Proof. This is straightforward with the interpretation by half cells of the proof of Proposition 2.27. Indeed applying the first-order scheme to the half cell values gives

$$\begin{aligned}
U^{n+1}_{i-1/4} &= U_{i-1/4} - \frac{2\Delta t}{\Delta x_i}\left(F_{i-} - F_{i-1/2+} \right), \\
U^{n+1}_{i+1/4} &= U_{i+1/4} - \frac{2\Delta t}{\Delta x_i}\left(F_{i+1/2-} - F_{i+} \right),
\end{aligned} \qquad (4.188)$$

with

$$F_{i-} = F_l\Big(U^n_{i-1/2+}, U^n_{i+1/2-}, Z^n_{i-1/2+}, Z^n_{i+1/2-}\Big),$$
$$F_{i+} = F_r\Big(U^n_{i-1/2+}, U^n_{i+1/2-}, Z^n_{i-1/2+}, Z^n_{i+1/2-}\Big). \tag{4.189}$$

By summing the two equations in (4.188) and using (2.230), we conclude that $U^{n+1}_i = (U^{n+1}_{i-1/4} + U^{n+1}_{i+1/4})/2$ if $F_c = F_r - F_l$, which gives the result. □

In practice, the choice $F_r - F_l$ cannot be used, because (4.187) is really necessary. A slight restriction on the reconstruction is also necessary, which is that whenever the sequence (U_i, Z_i) is realized as the cell averages of smooth functions $(U(x), Z(x))$, then

$$Z_{i+1/2+} - Z_{i+1/2-} = O\Big((\Delta x_i + \Delta x_{i+1})h\Big),$$
$$Z_{i+1/2-} - Z_{i-1/2+} = O(\Delta x_i). \tag{4.190}$$

These assumptions are very weak and of local nature, since in general one would have according to the definition of a second-order reconstruction operator errors in h^2 and h respectively on the right-hand sides of (4.190). These conditions are satisfied for example for the minmod reconstruction.

Proposition 4.18. *Let us assume that the numerical fluxes F_l, F_r are consistent in the sense of Definition 4.2, that F_l, F_r are Lipschitz continuous, that $F_r - F_l \in C^1$ with $d(F_r - F_l)$ Lipschitz continuous, that F_c satisfies (4.187) and that the reconstruction operator satisfies (4.190).*
If for all i,

$$U^n_i = \frac{1}{\Delta x_i} \int_{C_i} U(t_n, x)\, dx, \qquad Z_i = \frac{1}{\Delta x_i} \int_{C_i} Z(x)\, dx, \tag{4.191}$$

for some smooth solution $(U(t,x), Z(x))$ to (3.3), then U^{n+1}_i defined by (4.183)–(4.184) satisfies for all i

$$U^{n+1}_i = \frac{1}{\Delta x_i} \int_{C_i} U(t_{n+1}, x)\, dx + \Delta t\left(\frac{1}{\Delta x_i}(\mathcal{F}_{i+1/2} - \mathcal{F}_{i-1/2}) + \mathcal{E}_i\right), \tag{4.192}$$

where

$$\mathcal{F}_{i+1/2} = O(\Delta t) + O(h^2), \qquad \mathcal{E}_i = O(\Delta t) + O(h^2), \tag{4.193}$$

as Δt and $h = \sup_i \Delta x_i$ tend to 0.

Proof. Following the proof of Proposition 4.3, we have that (4.22) holds with (4.23) and (4.24). We use again the definitions (4.25), (4.26) of $Z_{i+1/2}$, $B_{i+1/2}$, B_i, and define the mean flux $F_{i+1/2}$ by

$$F_{i+1/2} = \frac{\Delta x_{i+1} F_{i+1/2-} + \Delta x_i F_{i+1/2+}}{\Delta x_i + \Delta x_{i+1}}$$
$$+ B_{i+1/2}\left(\frac{\Delta x_{i+1} Z_{i+1/2-} + \Delta x_i Z_{i+1/2+}}{\Delta x_i + \Delta x_{i+1}} - Z_{i+1/2}\right). \tag{4.194}$$

Notice that by the definition of a second-order reconstruction, we have $Z_{i+1/2-} = Z_{i+1/2}+O(h^2)$, $Z_{i+1/2+} = Z_{i+1/2}+O(h^2)$, thus the second line of (4.194) is $O(h^2)$. Then, we can rewrite (4.183) as (4.28), with

$$
\begin{aligned}
E_i &= \frac{1}{\Delta x_i}(F_{i+1/2-} - F_{i+1/2} + F_{i-1/2} - F_{i-1/2+} - \delta F_i) \\
&= -\frac{\delta F_i}{\Delta x_i} + \frac{F_{i+1/2-} - F_{i+1/2+}}{\Delta x_i + \Delta x_{i+1}} + \frac{F_{i-1/2-} - F_{i-1/2+}}{\Delta x_{i-1} + \Delta x_i} \\
&\quad - \frac{B_{i+1/2}}{\Delta x_i}\left(\frac{\Delta x_{i+1}Z_{i+1/2-} + \Delta x_i Z_{i+1/2+}}{\Delta x_i + \Delta x_{i+1}} - Z_{i+1/2}\right) \\
&\quad + \frac{B_{i-1/2}}{\Delta x_i}\left(\frac{\Delta x_i Z_{i-1/2-} + \Delta x_{i-1}Z_{i-1/2+}}{\Delta x_{i-1} + \Delta x_i} - Z_{i-1/2}\right).
\end{aligned}
\tag{4.195}
$$

Therefore, by subtracting (4.22) to (4.28), we get (4.192) with

$$
\mathcal{F}_{i+1/2} = \underline{F}_{i+1/2} - F_{i+1/2}, \qquad \mathcal{E}_i = \underline{E}_i - E_i. \tag{4.196}
$$

We observe that (4.31) is still valid, and that from (4.13) and the Lipschitz continuity of F_l, F_r,

$$
F_{i+1/2} = F\Big(U(t_n, x_{i+1/2}), Z(x_{i+1/2})\Big) + O(h^2), \tag{4.197}
$$

giving $\mathcal{F}_{i+1/2} = O(\Delta t) + O(h^2)$. Then, since $F_r - F_l \in C^{1,1}$, as in Remark 4.5 and because of (4.18), (4.14) holds with an error in $|Z_r - Z_l|(|U_l - U| + |U_r - U| + |Z_l - Z| + |Z_r - Z|)$. Thus

$$
\begin{aligned}
F_{i+1/2+} - F_{i+1/2-} &= -B_{i+1/2}\Delta Z_{i+1/2} + O\Big(h^2\Delta Z_{i+1/2}\Big) \\
&= -B_i\Delta Z_{i+1/2} + O\Big(h\Delta Z_{i+1/2}\Big),
\end{aligned}
\tag{4.198}
$$

and

$$
\begin{aligned}
F_{i-1/2+} - F_{i-1/2-} &= -B_{i-1/2}\Delta Z_{i-1/2} + O\Big(h^2\Delta Z_{i-1/2}\Big) \\
&= -B_i\Delta Z_{i-1/2} + O\Big(h\Delta Z_{i-1/2}\Big).
\end{aligned}
\tag{4.199}
$$

But by (4.187) and the second line of (4.190), we have

$$
\begin{aligned}
-\frac{\delta F_i}{\Delta x_i} &= \left(B\left(\frac{U_{i-1/2+} + U_{i+1/2-}}{2}, \frac{Z_{i-1/2+} + Z_{i+1/2-}}{2}\right) + O(h^2)\right)\frac{\Delta Z_i}{\Delta x_i} \\
&= (B_i + O(h^2))\frac{\Delta Z_i}{\Delta x_i} \\
&= B_i\frac{\Delta Z_i}{\Delta x_i} + O(h^2).
\end{aligned}
\tag{4.200}
$$

Therefore, reporting (4.198), (4.199) and (4.200) in (4.195) gives with the first line of (4.190)

$$
\begin{aligned}
E_i &= B_i \frac{\Delta Z_i}{\Delta x_i} + B_i \frac{\Delta Z_{i+1/2}}{\Delta x_i + \Delta x_{i+1}} + B_i \frac{\Delta Z_{i-1/2}}{\Delta x_{i-1} + \Delta x_i} \\
&\quad - \frac{B_i}{\Delta x_i} \left(\frac{\Delta x_{i+1} Z_{i+1/2-} + \Delta x_i Z_{i+1/2+}}{\Delta x_i + \Delta x_{i+1}} - Z_{i+1/2} \right) \\
&\quad + \frac{B_i}{\Delta x_i} \left(\frac{\Delta x_i Z_{i-1/2-} + \Delta x_{i-1} Z_{i-1/2+}}{\Delta x_{i-1} + \Delta x_i} - Z_{i-1/2} \right) + O(h^2) \\
&= B_i \frac{Z_{i+1/2} - Z_{i-1/2}}{\Delta x_i} + O(h^2) \\
&= B_i Z_x(x_i) + O(h^2).
\end{aligned}
\tag{4.201}
$$

Since

$$
\underline{E}_i = B_i Z_x(x_i) + O(\Delta t) + O(h^2),
\tag{4.202}
$$

we conclude that $\mathcal{E}_i = O(\Delta t) + O(h^2)$. $\qquad\square$

4.13.2 Well-balancing

The well-balancing property can be achieved if all the ingredients in the scheme (4.183)–(4.184) satisfy a suitable condition. We assume here that a family of discrete steady states is chosen, as described in Section 4.1.

Definition 4.19. *We say that the second-order reconstruction operator is well-balanced if to a sequence (U_i, Z_i) that is a discrete steady state, it associates a sequence of interface values*

$$
\ldots, \ (U, Z)_{i-1/2-}, \ (U, Z)_{i-1/2+}, \ (U, Z)_{i+1/2-}, \ (U, Z)_{i+1/2+}, \ \ldots
\tag{4.203}
$$

that is again a discrete steady state.

We remark that here it is important to consider global steady states sequences, since the reconstruction operator can have a large dependency stencil.

Proposition 4.20. *Assume that the reconstruction operator is well-balanced, that the numerical fluxes F_l, F_r are well-balanced in the sense of Definition 4.1, and that the centered flux F_c satisfies that whenever (U_l, U_r, Z_l, Z_r) is a local steady state,*

$$
F_c(U_l, U_r, Z_l, Z_r) = F(U_r, Z_r) - F(U_l, Z_l).
\tag{4.204}
$$

Then the second-order scheme (4.183)–(4.184) is well-balanced, in the sense that steady states sequences are let invariant.

Proof. This is obvious by the formulas (4.183)–(4.184). $\qquad\square$

4.13.3 Centered flux

The centered flux F_c has to satisfy (4.187) for second-order accuracy, and (4.204) for well-balancing. For Example 4.1 and if $F'(U) > 0$, we can take

$$F_c(U_l, U_r, \Delta Z) = -\frac{F(U_r) - F(U_l)}{D(U_r) - D(U_l)} \Delta Z. \tag{4.205}$$

For the Saint Venant system and if we retain only steady states at rest, we have a similar choice

$$F_c(U_l, U_r, \Delta Z) = \left(0, -\rho^* \Delta Z\right), \tag{4.206}$$

$$\rho^* = \frac{p(\rho_r) - p(\rho_l)}{e(\rho_r) + p(\rho_r)/\rho_r - e(\rho_l) - p(\rho_l)/\rho_l}. \tag{4.207}$$

For the physical pressure law $p(\rho) = \kappa\rho^2$, this simplifies in $\rho^* = (\rho_l + \rho_r)/2$. An important point is that since the first component in (4.206) is 0, which is the same value as the first component of $F_r - F_l$ by (4.45), we have that the density obtained by (4.183)–(4.184) is the same as it would be with the choice $F_c = F_r - F_l$, and therefore by Lemma 4.17, this density is nonnegative under a half CFL condition.

For systems that can be treated by the hydrostatic reconstruction method of Section 4.11, there is indeed a general procedure to derive a centered flux. The idea is to modify the choice $F_r - F_l$ in such a way that we keep the well-balanced property, while recentering the variables. Let us illustrate this with the case of the Saint Venant system with variable pressure of Section 4.11.1. We take the difference in (4.147),

$$F_c(U_l, U_r, Z_l, Z_r) = \left(0, p(\rho_r)\iota_r - p(\rho_r^*)\iota^* - p(\rho_l)\iota_l + p(\rho_l^*)\iota^*\right), \tag{4.208}$$

we keep (4.149),

$$\begin{aligned}
(e + p/\rho)(\rho_l^*) &= \left((e + p/\rho)(\rho_l)\iota_l + gz_l - gz^*\right)_+/\iota^*, \\
(e + p/\rho)(\rho_r^*) &= \left((e + p/\rho)(\rho_r)\iota_r + gz_r - gz^*\right)_+/\iota^*,
\end{aligned} \tag{4.209}$$

but (4.150) is replaced by a centered choice of z^* and ι^*, for example

$$z^* = \frac{z_l + z_r}{2}, \qquad \iota^* = \sqrt{\iota_l \iota_r}. \tag{4.210}$$

Then one can check that this centered flux is well-balanced and satisfies the centering requirement (4.187). Since the first component of (4.208) vanishes, the centering does not destroy the nonnegativity property. A variant with cutoff is also possible if in this centered flux (that depends only on Δz and not independently of z_l and z_r), we replace $g\Delta z$ by its projection onto the interval $[-2(e + p/\rho)(\rho_r)\iota_r, 2(e + p/\rho)(\rho_l)\iota_l]$.

4.13.4 Reconstruction operator

A second-order reconstruction for the Saint Venant system can be performed on the vector $(U_i, Z_i) = (\rho_i, \rho_i u_i, Z_i)$ as follows. We use the reconstruction proposed in Example 2.14 for U, and for the variable Z, we perform the minmod reconstruction on $\zeta = e(\rho) + p(\rho)/\rho + Z$. This means that we set $\zeta_i = e(\rho_i) + p(\rho_i)/\rho_i + Z_i$, from these values we compute $\zeta_{i+1/2\pm}$, and we finally set

$$Z_{i+1/2\pm} = \zeta_{i+1/2\pm} - \left(e(\rho_{i+1/2\pm}) + p(\rho_{i+1/2\pm})/\rho_{i+1/2\pm} \right). \qquad (4.211)$$

This reconstruction is obviously well-balanced with respect to steady states at rest. Consequently, according to Proposition 4.20, with this choice of second-order reconstruction and with F_c defined by (4.206)–(4.207), the whole second-order scheme is well-balanced. This choice of reconstructing ζ has the advantage to treat correctly interfaces between dry and wet cells, contrarily to other choices, see [6].

Chapter 5

Multidimensional finite volumes with sources

We do not want to give here a full description of finite volume methods in multidimension, but rather to explain a few topics especially related to source terms, following [13], [7], [75]. Even if our considerations apply to any dimension, our presentation is restricted for simplicity to the two-dimensional case.

We are going to consider in this chapter a multidimensional system of the form

$$\partial_t U + \partial_1(F_1(U, Z)) + \partial_2(F_2(U, Z)) + B_1(U, Z)\partial_1 Z + B_2(U, Z)\partial_2 Z = 0, \quad (5.1)$$

where the space variable is $x = (x_1, x_2) \in \mathbb{R}^2$, and the partial derivatives ∂_1, ∂_2 refer to these variables. The unknown is $U(t, x) \in \mathbb{R}^p$, $Z = Z(x) \in \mathbb{R}^r$ is given, and the nonlinearities F_1, F_2, B_1, B_2 are smooth. This system is the natural generalization of (3.1) to two dimensions, and can be written as a quasilinear system in multidimension for the variable $\widetilde{U} = (U, Z)$,

$$\begin{cases} \partial_t U + (\partial_U F_1)(U, Z)\partial_1 U + (\partial_U F_2)(U, Z)\partial_2 U \\ \quad + ((\partial_Z F_1)(U, Z) + B_1(U, Z)) \partial_1 Z + ((\partial_Z F_2)(U, Z) + B_2(U, Z)) \partial_2 Z = 0, \\ \partial_t Z = 0. \end{cases}$$

$$(5.2)$$

For a *two-dimensional quasilinear system*

$$\partial_t \widetilde{U} + A_1(\widetilde{U})\partial_1 \widetilde{U} + A_2(\widetilde{U})\partial_2 \widetilde{U} = 0, \quad (5.3)$$

we can consider planar solutions, i.e. solutions of the form $\widetilde{U}(t, x) = \widetilde{U}(t, \zeta)$ with $\zeta = x \cdot n$ and $n = (n^1, n^2)$ is any unit vector in $\mathbb{R}^2$, which leads to

$$\partial_t \widetilde{U} + A_n(\widetilde{U})\partial_\zeta \widetilde{U} = 0, \quad (5.4)$$

with

$$A_n(\widetilde{U}) = n^1 A_1(\widetilde{U}) + n^2 A_2(\widetilde{U}). \quad (5.5)$$

The notions introduced in Chapter 1 can be applied to the one-dimensional quasilinear system (5.4), and by this way one defines hyperbolicity, entropies, and other notions for (5.3) by requiring the one-dimensional properties for all directions n. In particular, the system (5.3) is hyperbolic if for any unit vector n in $\mathbb{R}^2$ and any $\widetilde{U}$, $A_n(\widetilde{U})$ is diagonalizable. One should indeed also require a smooth dependence with respect to n of the eigenvalues and eigenvectors, but we shall not enter into details here, the reader is referred to the literature, for example [91].

5.1   Nonconservative finite volumes

The finite volume method for solving multidimensional systems is described in detail in [33] for example. It involves a mesh made of cells C_i, the *control volumes*. The interface Γ_{ij} between two cells C_i and C_j is assumed to be a subset of an hyperplane, thus the unit normal n_{ij} oriented from C_i to C_j is well-defined (see Figure 5.1). In particular, $n_{ji} = -n_{ij}$.

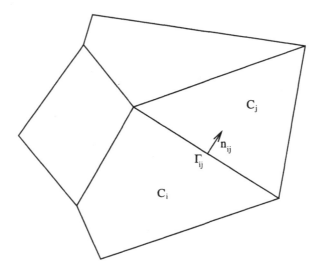

Figure 5.1: Interface Γ_{ij} between cells C_i and C_j

A finite volume method for solving (5.1) is a formula for updating the values U_i^n attached to each cell C_i, of the form

$$U_i^{n+1} - U_i + \frac{\Delta t}{|C_i|} \sum_{j \in K_i} |\Gamma_{ij}| F_{ij} = 0, \tag{5.6}$$

where as before, U_i stands for U_i^n, Δt is the timestep, $|C_i|$ is the volume of C_i, $|\Gamma_{ij}|$ is the length of Γ_{ij}, K_i is the set of indices j corresponding to cells C_j having a common interface with C_i, and F_{ij} is an exchange term between C_j and C_i. This formula generalizes (4.1). When (5.1) is conservative, i.e. $B_1 = B_2 \equiv 0$, it is natural to ask that (5.6) is also conservative, i.e. $F_{ji} = -F_{ij}$, that ensures that $\sum_i |C_i| U_i^n$ is time independent.

In our case, there are nonconservative terms in (5.1), thus we have to take this into account, by generalizing the nonconservative schemes of Chapter 4. Therefore, we take the interface fluxes of the form

$$F_{ij} = F(U_i, U_j, Z_i, Z_j, n_{ij}), \tag{5.7}$$

where $F(U_l, U_r, Z_l, Z_r, n)$ is a numerical flux approximating $n^1 F_1 + n^2 F_2$, and the indices l and r are related to the orientation of n. To relate this notation with the one-dimensional one, one has to understand this numerical flux as the value *on the left of the interface*, knowing that n is going from left to right. The value on the right is then $-F(U_r, U_l, Z_r, Z_l, -n)$.

5.2 Well-balancing

The above formulation means indeed that we are solving the problem (5.1) *by interface*. It follows that the steady states that can be preserved by such a method are those that can be seen as interface by interface steady states. In particular, more general steady states where the dependence in x_1 and x_2 are nontrivially balanced cannot be preserved. Discrete approximations of the steady states are selected via an interface relation

$$\mathcal{D}(U_l, U_r, Z_l, Z_r, n) = 0, \tag{5.8}$$

and we shall assume for coherence that if (U_l, U_r, Z_l, Z_r, n) satisfies this relation, then $(U_r, U_l, Z_r, Z_l, -n)$ also. The well-balanced property of Definition 4.1 generalizes as follows.

Definition 5.1. *The scheme* (5.6)–(5.7) *is well-balanced relatively to some family of discrete steady state defined by* (5.8) *if one has for any data satisfying* (5.8)

$$F(U_l, U_r, Z_l, Z_r, n) = n^1 F_1(U_l, Z_l) + n^2 F_2(U_l, Z_l). \tag{5.9}$$

According to (5.6)–(5.7), this property guarantees that if at time t_n we start with a sequence (U_i) that is a discrete steady state at each interface, then it remains unchanged at the next time level, because by the Stokes formula,

$$\sum_{j \in K_i} |\Gamma_{ij}| n_{ij} = \int_{\partial C_i} n = 0. \tag{5.10}$$

5.3 Consistency

The consistency Definition 4.2 generalizes as follows.

Definition 5.2. *We say that the scheme* (5.6)–(5.7) *is consistent with* (5.1) *if the numerical flux satisfies the consistency with the exact flux*

$$F(U, U, Z, Z, n) = n^1 F_1(U, Z) + n^2 F_2(U, Z) \text{ for any } U, Z, n, \tag{5.11}$$

and the asymptotic conservativity/consistency with the source

$$\begin{aligned} F(U_l, U_r, Z_l, Z_r, n) + F(U_r, U_l, Z_r, Z_l, -n) \\ = (n^1 B_1(U, Z) + n^2 B_2(U, Z))(Z_r - Z_l) + o(Z_r - Z_l), \end{aligned} \tag{5.12}$$

as $U_l, U_r \to U$ *and* $Z_l, Z_r \to Z$.

Note that the property (5.12) implies in particular that $F(U_l, U_r, Z, Z, n) + F(U_r, U_l, Z, Z, -n) = 0$, which is the conservativity when Z is constant. The previous definition is justified by the following result, formulated in the multidimensional weak sense.

Proposition 5.3. *Let $x_i \in C_i$ be chosen arbitrarily, and assume that for all i,*

$$U_i^n = \frac{1}{|C_i|} \int_{C_i} U(t_n, x)\, dx, \qquad Z_i = Z(x_i) + O(diam(C_i)), \qquad (5.13)$$

for some smooth solution $(U(t, x), Z(x))$ to (5.1), and define U_i^{n+1} by (5.6)–(5.7). If the numerical flux is consistent and if

$$\frac{diam(C_i)}{|C_i|} \sum_{j \in K_i} |\Gamma_{ij}| \ is\ bounded, \qquad (5.14)$$

then for all i,

$$U_i^{n+1} = \frac{1}{|C_i|} \int_{C_i} U(t_{n+1}, x)\, dx + \Delta t \left(\frac{1}{|C_i|} \sum_{j \in K_i} |\Gamma_{ij}| \mathcal{F}_{ij} + \mathcal{E}_i \right), \qquad (5.15)$$

where

$$\mathcal{F}_{ji} = -\mathcal{F}_{ij}, \qquad (5.16)$$

and

$$\mathcal{F}_{ij} \to 0, \qquad \mathcal{E}_i \to 0, \qquad (5.17)$$

as Δt and $h \equiv \sup_i diam(C_i)$ tend to 0.

Proof. Integrate the equation (5.1) satisfied by $U(t, x)$ with respect to t and x over $]t_n, t_{n+1}[\times C_i$, and divide the result by $|C_i|$. According to the Stokes formula, we get

$$\frac{1}{|C_i|} \int_{C_i} U(t_{n+1}, x)dx - \frac{1}{|C_i|} \int_{C_i} U(t_n, x)dx + \frac{\Delta t}{|C_i|} \sum_{j \in K_i} |\Gamma_{ij}| \underline{F}_{ij} + \Delta t \underline{E}_i = 0,$$

$$(5.18)$$

where $\underline{F}_{ij}$ is the exact interface flux

$$\underline{F}_{ij} = \frac{1}{\Delta t} \int_{t_n}^{t_{n+1}} \frac{1}{|\Gamma_{ij}|} \int_{\Gamma_{ij}} \left(n_{ij}^1 F_1 + n_{ij}^2 F_2 \right) (U(t, x), Z(x))\, dx dt, \qquad (5.19)$$

and $\underline{E}_i$ is the exact mean source

$$\underline{E}_i = \frac{1}{\Delta t} \int_{t_n}^{t_{n+1}} \frac{1}{|C_i|} \int_{C_i} \left(B_1(U, Z)\partial_1 Z + B_2(U, Z)\partial_2 Z \right) (t, x)\, dt dx. \qquad (5.20)$$

We have obviously $\underline{F}_{ji} = -\underline{F}_{ij}$. Now, denote by x_{ij} the middle of Γ_{ij}, and

$$Z_{ij} = Z(x_{ij}), \quad B_1^{ij} = B_1(U(t_n, x_{ij}), Z_{ij}), \quad B_2^{ij} = B_2(U(t_n, x_{ij}), Z_{ij}). \quad (5.21)$$

We take $\mu_{ij} \geq 0$, $\mu_{ij} + \mu_{ji} = 1$ (to be chosen later on), and define a mean flux

$$\widetilde{F}_{ij} = \mu_{ij} F_{ij} - \mu_{ji} F_{ji} + (n_{ij}^1 B_1^{ij} + n_{ij}^2 B_2^{ij})(\mu_{ij} Z_i + \mu_{ji} Z_j - Z_{ij}). \quad (5.22)$$

Then $\widetilde{F}_{ji} = -\widetilde{F}_{ij}$, and we can rewrite (5.6) as

$$U_i^{n+1} - U_i + \frac{\Delta t}{|C_i|} \sum_{j \in K_i} |\Gamma_{ij}| \widetilde{F}_{ij} + \Delta t \widetilde{E}_i = 0, \quad (5.23)$$

with

$$\widetilde{E}_i = \frac{1}{|C_i|} \sum_{j \in K_i} |\Gamma_{ij}| (F_{ij} - \widetilde{F}_{ij}). \quad (5.24)$$

Therefore, subtracting (5.18) to (5.23) gives (5.15) with

$$\mathcal{F}_{ij} = \underline{F}_{ij} - \widetilde{F}_{ij}, \qquad \mathcal{E}_i = \underline{E}_i - \widetilde{E}_i. \quad (5.25)$$

By difference, (5.16) is satisfied, and it only remains to prove (5.17). Noticing that by (5.13),

$$Z_j - Z_i = O(\text{diam}(C_i) + \text{diam}(C_j)), \quad (5.26)$$

we have by the consistency (5.11) (if the numerical flux is continuous)

$$\widetilde{F}_{ij} = n_{ij}^1 F_1(U(t_n, x_{ij}), Z_{ij}) + n_{ij}^2 F_2(U(t_n, x_{ij}), Z_{ij}) + o(1), \quad (5.27)$$

thus obviously from (5.19),

$$\mathcal{F}_{ij} = O(\Delta t) + o(1). \quad (5.28)$$

Now, let us make the choice

$$\mu_{ji} = \frac{\text{diam}(C_i)}{\text{diam}(C_i) + \text{diam}(C_j)}. \quad (5.29)$$

We compute from (5.22) and the consistency (5.12)

$$
\begin{aligned}
& F_{ij} - \widetilde{F}_{ij} \\
&= \mu_{ji}(F_{ij} + F_{ji}) - (n_{ij}^1 B_1^{ij} + n_{ij}^2 B_2^{ij})(\mu_{ij} Z_i + \mu_{ji} Z_j - Z_{ij}) \\
&= (n_{ij}^1 B_1^{ij} + n_{ij}^2 B_2^{ij})(Z_{ij} - Z_i) + o(\mu_{ji}(\text{diam}(C_i) + \text{diam}(C_j))) \\
&= (n_{ij}^1 B_1(U_i, Z_i) + n_{ij}^2 B_2(U_i, Z_i))(Z_{ij} - Z_i) + o(\text{diam}(C_i)).
\end{aligned}
\quad (5.30)
$$

Then, because of (5.10), when taking the sum (5.24), we can replace $Z_{ij} - Z_i$ in the last line of (5.30) by $Z_{ij} - Z(x_i) = \partial_x Z(x_i) \cdot (x_{ij} - x_i) + o(\text{diam}(C_i))$. Therefore, we obtain

$$\widetilde{E}_i = \frac{1}{|C_i|} \sum_{j \in K_i} |\Gamma_{ij}| \left[(n_{ij}^1 B_1(U_i, Z_i) + n_{ij}^2 B_2(U_i, Z_i)) \partial_x Z(x_i) \cdot (x_{ij} - x_i) \right.$$

$$\left. + o(\text{diam}(C_i)) \right]$$

$$= B_1(U_i, Z_i) \partial_x Z(x_i) \cdot \frac{1}{|C_i|} \sum_{j \in K_i} |\Gamma_{ij}| n_{ij}^1 (x_{ij} - x_i) \tag{5.31}$$

$$+ B_2(U_i, Z_i) \partial_x Z(x_i) \cdot \frac{1}{|C_i|} \sum_{j \in K_i} |\Gamma_{ij}| n_{ij}^2 (x_{ij} - x_i)$$

$$+ o\left(\frac{\text{diam}(C_i)}{|C_i|} \sum_{j \in K_i} |\Gamma_{ij}| \right).$$

But according to the Stokes formula,

$$\frac{1}{|C_i|} \sum_{j \in K_i} |\Gamma_{ij}| n_{ij}^1 (x_{ij} - x_i) = \frac{1}{|C_i|} \int_{\partial C_i} n^1(x - x_i) dx = \frac{1}{|C_i|} \int_{C_i} \partial_1(x - x_i)\, dx = (1, 0), \tag{5.32}$$

and a similar formula holds for the other sum. Therefore, with (5.14) we get

$$\widetilde{E}_i = B_1(U_i, Z_i) \partial_1 Z(x_i) + B_2(U_i, Z_i) \partial_2 Z(x_i) + o(1), \tag{5.33}$$

and this gives obviously that $\mathcal{E}_i = o(1)$. $\qquad\qquad\qquad\qquad\qquad\qquad\square$

Remark 5.1. Second-order methods are also possible with sources, see [8].

Invariant domains and entropy inequalities can be described very similar to the one-dimensional case, and incell properties are deduced again from interface properties at the price of diminishing the CFL condition. However we shall not describe this here.

5.4 Additional source terms

A generalization of the approach proposed in Section 4.12 is as follows. Consider a system of the form (5.1), but where F_1, F_2, B_1, B_2 are independent of Z,

$$\partial_t U + \partial_1 (F_1(U)) + \partial_2 (F_2(U)) + B_1(U) \partial_1 Z + B_2(U) \partial_2 Z = 0. \tag{5.34}$$

This problem is invariant under addition of a constant to Z. Then, assume known a numerical flux which, coherently with this property, does not depend arbitrarily on Z_l and Z_r, but only on the difference $\Delta Z = Z_r - Z_l$,

$$F(U_l, U_r, Z_l, Z_r, n) = F(U_l, U_r, \Delta Z, n). \tag{5.35}$$

Assume then that one would like to solve a problem (5.34) with an additional zero-order source of the form

$$\partial_t U + \partial_1(F_1(U)) + \partial_2(F_2(U)) + B_1(U)\partial_1 Z + B_2(U)\partial_2 Z$$
$$= B_1(U)f_1(t,x) + B_2(U)f_2(t,x). \tag{5.36}$$

A finite volume method for solving (5.36) is

$$U_i^{n+1} - U_i + \frac{\Delta t}{|C_i|} \sum_{j \in K_i} |\Gamma_{ij}| F_{ij} = 0, \tag{5.37}$$

with

$$F_{ij} = F\left(U_i, U_j, \Delta Z_{ij} - f_1^{ij}(x_j - x_i)_1 - f_2^{ij}(x_j - x_i)_2, n_{ij}\right), \tag{5.38}$$

where $\Delta Z_{ij} = Z_j - Z_i$, $x_i \in C_i$ is chosen arbitrarily (independently of the interface Γ_{ij}), and f_1^{ij}, f_2^{ij} are approximate values of f_1, f_2 on Γ_{ij}, with $f^{ji} = f^{ij}$. By looking at the proof of Proposition 5.3, the scheme (5.37)–(5.38) is easily seen to be consistent with (5.36) (in a sense that we shall not write down here), and as in the one-dimensional case it also inherits the well-balanced property of the original scheme.

5.5 Two-dimensional Saint Venant system

Let us illustrate the notions introduced above with the two-dimensional Saint Venant system

$$\begin{cases} \partial_t \rho + \partial_1(\rho u) + \partial_2(\rho v) = 0, \\ \partial_t(\rho u) + \partial_1(\rho u^2 + p(\rho)) + \partial_2(\rho u v) + \rho g \partial_1 z = 0, \\ \partial_t(\rho v) + \partial_1(\rho u v) + \partial_2(\rho v^2 + p(\rho)) + \rho g \partial_2 z = 0, \end{cases} \tag{5.39}$$

where $\rho(t,x) \geq 0$ is the water height, $u(t,x)$, $v(t,x)$ are the two components of the velocity field, the pressure $p(\rho)$ satisfies as usual $p'(\rho) > 0$ (the physically relevant case being $p(\rho) = g\rho^2/2$), $g > 0$ is the gravity constant, and $z(x)$ is the topography. We define as in one dimension

$$Z = gz, \tag{5.40}$$

so that the system can be written as (5.1) with

$$U = \begin{pmatrix} \rho \\ \rho u \\ \rho v \end{pmatrix}, \quad F_1 = \begin{pmatrix} \rho u \\ \rho u^2 + p(\rho) \\ \rho u v \end{pmatrix}, \quad F_2 = \begin{pmatrix} \rho v \\ \rho u v \\ \rho v^2 + p(\rho) \end{pmatrix}, \tag{5.41}$$

and

$$B_1 = \begin{pmatrix} 0 \\ \rho \\ 0 \end{pmatrix}, \qquad B_2 = \begin{pmatrix} 0 \\ 0 \\ \rho \end{pmatrix}. \tag{5.42}$$

The steady states at rest are characterized by

$$u = v = 0, \qquad e(\rho) + \frac{p(\rho)}{\rho} + Z = cst, \tag{5.43}$$

with $e'(\rho) = p(\rho)/\rho^2$.

A property of this system is that it is invariant under rotation. It can be seen as follows. Let $n = (n^1, n^2)$ be a unit vector, and

$$R_n = \begin{pmatrix} n^1 & -n^2 \\ n^2 & n^1 \end{pmatrix} \tag{5.44}$$

its associated rotation. If

$$x = R_n x', \qquad (u', v') = R_n^{-1}(u, v), \qquad U' = (\rho, \rho u', \rho v'), \tag{5.45}$$

then U' is again a solution as a function of (t, x'). Indeed the flux in the direction of n is

$$n^1 F_1(U) + n^2 F_2(U) = \begin{pmatrix} \rho u' \\ R_n \begin{pmatrix} \rho u'^2 + p(\rho) \\ \rho u' v' \end{pmatrix} \end{pmatrix}. \tag{5.46}$$

This rotational invariance enables to define a numerical flux from one-dimensional numerical fluxes $F_l(U_l, U_r, \Delta Z)$, $F_r(U_l, U_r, \Delta Z)$ by the formula

$$F(U_l, U_r, \Delta Z, n) = \begin{pmatrix} F_l^0(U_l', U_r', \Delta Z) \\ R_n \begin{pmatrix} F_l^1(U_l', U_r', \Delta Z) \\ F_l^2(U_l', U_r', \Delta Z) \end{pmatrix} \end{pmatrix}. \tag{5.47}$$

The one-dimensional numerical fluxes $F_l(U_l, U_r, \Delta Z)$, $F_r(U_l, U_r, \Delta Z)$ are associated to the problem

$$\begin{cases} \partial_t \rho + \partial_x(\rho u) = 0, \\ \partial_t(\rho u) + \partial_x(\rho u^2 + p(\rho)) + \rho \partial_x Z = 0, \\ \partial_t(\rho v) + \partial_x(\rho u v) = 0, \end{cases} \tag{5.48}$$

and one has to impose the symmetry by changing x to $-x$ and u in $-u$, v in $-v$,

$$F_r(U_l, U_r, \Delta Z) = -F_l(U_r^\sharp, U_l^\sharp, -\Delta Z)^\sharp, \tag{5.49}$$

with $(\rho, \rho u, \rho v)^\sharp \equiv (\rho, -\rho u, -\rho v)$. This implies the formula associated to (5.47),

$$-F(U_r, U_l, -\Delta Z, -n) = \begin{pmatrix} F_r^0(U_l', U_r', \Delta Z) \\ R_n \begin{pmatrix} F_r^1(U_l', U_r', \Delta Z) \\ F_r^2(U_l', U_r', \Delta Z) \end{pmatrix} \end{pmatrix}. \tag{5.50}$$

The following result can be checked easily.

Proposition 5.4. *If F_l, F_r are consistent numerical fluxes for (5.48) that satisfy (5.49), then the numerical flux (5.47) is consistent with (5.39). Moreover, if F_l, F_r are well-balanced with respect to the discrete steady states at rest, then (5.47) also.*

Notice that the last equation in (5.48) is a passive transport equation, and can be solved according to Section 2.7.

To illustrate the additional source approach of Section 5.4, consider the Saint Venant system with force

$$\begin{cases} \partial_t \rho + \partial_1(\rho u) + \partial_2(\rho v) = 0, \\ \partial_t(\rho u) + \partial_1(\rho u^2 + p(\rho)) + \partial_2(\rho u v) + \rho \partial_1 Z = \rho f_1, \\ \partial_t(\rho v) + \partial_1(\rho u v) + \partial_2(\rho v^2 + p(\rho)) + \rho \partial_2 Z = \rho f_2. \end{cases} \tag{5.51}$$

It is of the form (5.36), thus we can treat $f = (f_1, f_2)$ by (5.38), one just has to define an interface value $f^{ij} = f^{ji}$ and use the argument $\Delta Z_{ij} - f^{ij} \cdot (x_j - x_i)$ in the numerical flux. This can be done for example in the case of Coriolis force $f = \omega(x)(-v, u)$, by taking $f^{ij} = \omega^{ij}(-(v_i + v_j)/2, (u_i + u_j)/2)$, see [19], [75], where a cartesian mesh is used. Another example is the Coulomb friction in two dimensions,

$$f = -g\mu \frac{(u, v)}{\sqrt{u^2 + v^2}}, \tag{5.52}$$

with $\mu \geq 0$, and the ratio is multivalued for $(u, v) = 0$, as in Section 4.12.1. In this case, we can take

$$f^{ij} = -\varphi_{g\mu} \left((e(\rho_i) + p(\rho_i)/\rho_i - e(\rho_j) - p(\rho_j)/\rho_j - \Delta Z_{ij}) \frac{x_j - x_i}{|x_j - x_i|^2}, \frac{(u^{ij}, v^{ij})}{\Delta t} \right), \tag{5.53}$$

with (4.182), (4.176), and for example

$$u^{ij} = \frac{\rho_i u_i + \rho_j u_j}{\rho_i + \rho_j}, \qquad v^{ij} = \frac{\rho_i v_i + \rho_j v_j}{\rho_i + \rho_j}. \tag{5.54}$$

Numerical results with this method can be found in [83].

Chapter 6

Numerical tests with source

For all tests in this section we take $p(\rho) = g\rho^2/2$ with $g = 9.81$ in the Saint Venant system.

Test 1: Accuracy test

This test is designed to evaluate the accuracy of the schemes. The solutions are continuous but have discontinuities in derivatives. The space variable x is taken in $[0, 40]$, the topography is

$$z(x) = \begin{cases} 0.48 \left(1 - \left(\dfrac{x - 20}{4}\right)^2\right) & \text{if } |x - 20| \leq 4, \\ 0 & \text{otherwise,} \end{cases} \tag{6.1}$$

initial data are $\rho^0 = 4$, $u^0 = 10/4$, and the final time is $t = 1$. We test the Suliciu method of Section 4.7, the kinetic method of Section 4.8, the VFRoe method of Section 4.9, and the hydrostatic reconstruction method of Section 4.11. This last method is used in conjunction with the Suliciu solver for the problem without source, but other solvers give very similar results. A reference solution is computed by using a very fine mesh of 3000 points. Table 6.1 shows the L^1 error and the numerical order of accuracy for first-order computations. The CFL number is 1, except for VFRoe where it is 0.99. Figures 6.1, 6.2, 6.3, 6.4, 6.5 represent respectively the water level $\rho + z$, the velocity u, the discharge ρu, the second 0-Riemann invariant $u^2/2 + g(\rho + z)$, and the Froude number $u/\sqrt{p'(\rho)}$ for 50 points. Tables 6.2 and 6.3 show the same diagnostics for second-order in time and space, with the reconstruction of Section 4.13.4 and respectively minmod and ENOm slope limiters. The CFL condition is taken half of the one used at first-order.

Cells	Suliciu		Kinetic		VFRoe		Hydrostatic	
50	3.00E-0	/	4.73E-0	/	2.72E-0	/	3.75E-0	/
100	1.63E-0	0.88	2.73E-0	0.79	1.48E-0	0.88	2.02E-0	0.89
200	8.25E-1	0.98	1.51E-0	0.85	8.09E-1	0.87	1.07E-0	0.92
400	4.30E-1	0.94	8.25E-1	0.87	4.37E-1	0.89	5.61E-1	0.93
800	2.25E-1	0.93	4.40E-1	0.91	2.32E-1	0.91	2.92E-1	0.94

Table 6.1: L^1 error and numerical order of accuracy for Test 1, first-order

The four methods give comparable results, and they are even more similar at second-order. The numerical order of accuracy for first-order is almost optimal, but at second-order it is limited by the presence of the discontinuities in the derivatives. The ENO limiter improves the resolution by a factor say approximately 0.6 with respect to the minmod limiter.

Cells	Suliciu		Kinetic		VFRoe		Hydrostatic	
50	3.03E-0	/	3.20E-0	/	3.01E-0	/	3.20E-0	/
100	1.18E-0	1.36	1.24E-0	1.37	1.18E-0	1.35	1.25E-0	1.36
200	4.69E-1	1.33	4.95E-1	1.32	4.69E-1	1.33	4.79E-1	1.38
400	1.90E-1	1.30	2.00E-1	1.31	1.90E-1	1.30	1.90E-1	1.33
800	7.39E-2	1.36	8.40E-2	1.25	7.39E-2	1.36	7.33E-2	1.37

Table 6.2: L^1 error and numerical order of accuracy for Test 1, second-order minmod

Cells	Suliciu		Kinetic		VFRoe		Hydrostatic	
50	1.98E-0	/	1.90E-0	/	1.97E-0	/	2.06E-0	/
100	8.60E-1	1.20	7.81E-1	1.28	8.63E-1	1.19	8.42E-1	1.29
200	3.14E-1	1.45	2.86E-1	1.45	3.13E-1	1.46	3.05E-1	1.47
400	1.21E-1	1.38	1.07E-1	1.42	1.20E-1	1.38	1.18E-1	1.37
800	4.24E-2	1.51	4.43E-2	1.27	4.24E-2	1.50	4.14E-2	1.51

Table 6.3: L^1 error and numerical order of accuracy for Test 1, second-order ENO

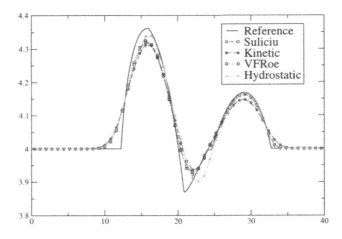

Figure 6.1: Water level for Test 1, first-order

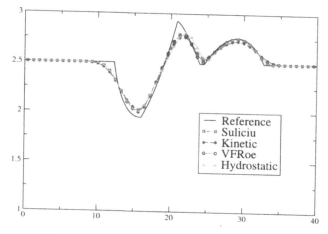

Figure 6.2: Velocity for Test 1, first-order

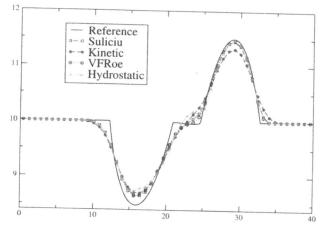

Figure 6.3: Discharge for Test 1, first-order

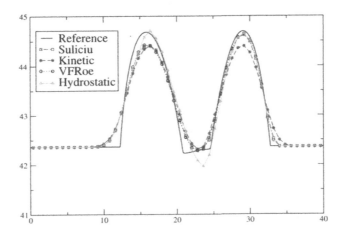

Figure 6.4: Riemann invariant for Test 1, first-order

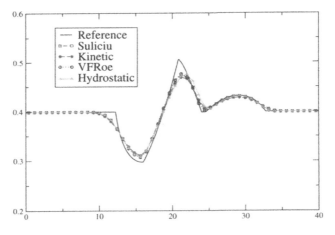

Figure 6.5: Froude number for Test 1, first-order

Test 2: Transcritical flow with shock

This test is taken from [38], [51], and is designed to assess the long time behavior and convergence to a steady state. The space domain is $[0, 25]$, the topography is

$$z(x) = \begin{cases} 0.2 - 0.05(x - 10)^2 & \text{if } 8 < x < 12, \\ 0 & \text{otherwise,} \end{cases} \tag{6.2}$$

the initial data are $\rho^0 = 0.33$, $u^0 = 0.18/0.33$, and boundary conditions are $\rho u(x = 0) = 0.18$, $\rho(x = 25) = 0.33$. The final time is $t = 200$.

The same four methods as above are tested, with a suitable treatment of the boundary conditions that we shall not describe here. We use 200 cells in space. Figure 6.6 shows the water level $\rho + z$ for first-order computations. The CFL condition is 1 except for VFRoe where it is 0.99, with the entropy fix of the problem without source. This entropy fix is not satisfactory since it gives oscillations here, and it gives a wrong solution for smaller CFL number, due to a nonentropy discontinuity arising at the critical point. According to the authors of [38], no efficient entropy fix has been found until now. A zoom of the interesting part is plotted on Figure 6.7, and the Discharge ρu is on Figure 6.8 (it should be constant in this test). The resolution of the Suliciu solver is a bit imprecise around the critical point, located at the point of maximum of z. The number of timesteps used is respectively 5572, 5601, 4932, 5068.

Figures 6.9 and 6.10 show the same test at second-order in time and space with ENOm slope, at CFL 0.5. We have not been able to obtain a result with the VFRoe solver. The other methods give similar results. We remark a small oscillation in front of the shock.

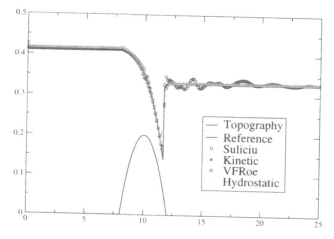

Figure 6.6: Water level for Test 2, first-order

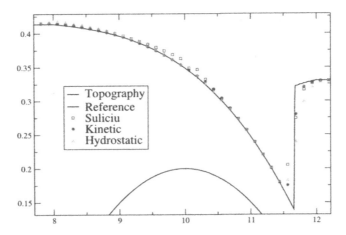

Figure 6.7: Zoom of water level for Test 2, first-order

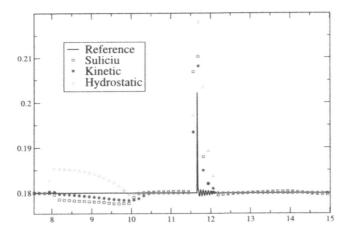

Figure 6.8: Discharge for Test 2, first-order

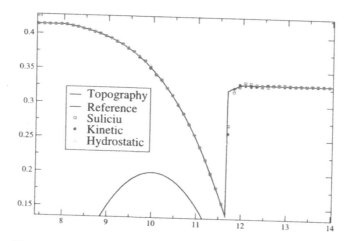

Figure 6.9: Zoom of water level for Test 2, second-order

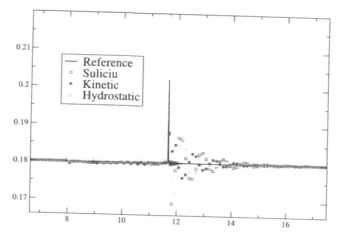

Figure 6.10: Discharge for Test 2, second-order

Test 3: Vacuum occurrence by a double rarefaction

This test is taken from [38], and shows the behavior with vacuum. The domain is again $[0, 25]$, the topography is

$$z(x) = \begin{cases} 1 & \text{if } 25/3 < x < 12.5, \\ 0 & \text{otherwise}, \end{cases} \tag{6.3}$$

and initial conditions are $\rho + z = 10$,

$$\rho(x)u(x) = \begin{cases} -350 & \text{if } x < 50/3, \\ 350 & \text{otherwise}. \end{cases} \tag{6.4}$$

Neumann boundary conditions are used, and the final time is $t = 0.25$.

The runs are performed with 200 points. Figures 6.11 and 6.12 show the water level for first-order and second-order schemes respectively, for the hydrostatic method with either the kinetic solver or the Suliciu solver as underlying homogeneous schemes. The CFL conditions are 1 at first-order, and 0.5 at second-order. The second-order computations take respectively 209 and 194 timesteps. We notice that as in the tests of Section 2.9, the second-order resolution has the defect to give too low densities around the vacuum. The Suliciu solver behaves a little better than the kinetic one on this test.

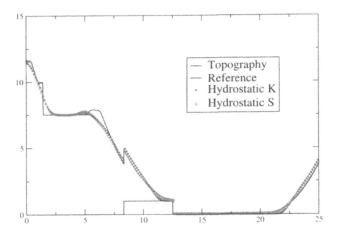

Figure 6.11: Water level for Test 3, first-order

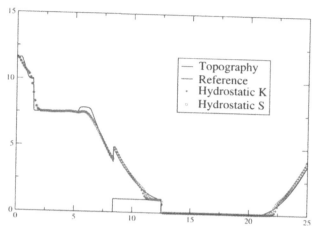

Figure 6.12: Water level for Test 3, second-order

Bibliography

[1] D. Amadori, L. Gosse, G. Guerra, *Global BV entropy solutions and uniqueness for hyperbolic systems of balance laws*, Arch. Ration. Mech. Anal. **162** (2002), 327-366.

[2] D. Amadori, L. Gosse, G. Guerra, *Godunov-type approximation for a general resonant balance law with large data*, J. Diff. Eq. **198** (2004), 233-274.

[3] N. Andrianov, *CONSTRUCT: a collection of MATLAB routines for constructing the exact solution to the Riemann problem for the shallow water equations*, available at http://www-ian.math.uni-magdeburg.de/home/andriano/CONSTRUCT.

[4] N. Andrianov, G. Warnecke, *On the solution to the Riemann problem for the compressible duct flow*, to appear in SIAM J. Appl. Math.

[5] D. Aregba-Driollet, R. Natalini, *Discrete kinetic schemes for multidimensional systems of conservation laws*, SIAM J. Numer. Anal. **37** (2000), 1973-2004.

[6] E. Audusse, F. Bouchut, M.-O. Bristeau, R. Klein, B. Perthame, *A fast and stable well-balanced scheme with hydrostatic reconstruction for shallow water flows*, to appear in SIAM J. Sci. Comp. (2004).

[7] E. Audusse, M.-O. Bristeau, B. Perthame, *Kinetic schemes for Saint Venant equations with source terms on unstructured grids*, INRIA Report RR-3989 (2000), http://www.inria.fr/rrrt/rr-3989.html

[8] E. Audusse, M.-O. Bristeau, B. Perthame, *Second order kinetic scheme for Saint Venant equations with source terms on unstructured grids*, preprint 2004.

[9] D.S. Bale, R.J. Leveque, S. Mitran, J.A. Rossmanith, *A wave propagation method for conservation laws and balance laws with spatially varying flux functions*, SIAM J. Sci. Comput. **24** (2002), 955-978.

[10] P. Batten, N. Clarke, C. Lambert, D.M. Causon, *On the choice of wavespeeds for the HLLC Riemann solver*, SIAM J. Sci. Comput. **18** (1997), 1553-1570.

[11] M. Baudin, C. Berthon, F. Coquel, P. Hoche, R. Masson, Q.H. Tran, *A relaxation method for two-phase flow models with hydrodynamic closure law*, preprint Univ Bordeaux (2002).

[12] A. Bermúdez, M.E. Vásquez, *Upwind methods for hyperbolic conservation laws with source terms*, Comput. Fluids **23** (1994), 1049-1071.

[13] A. Bermúdez, A. Dervieux, J.A. Desideri, M.E. Vásquez, *Upwind schemes for two-dimensional shallow water equations with variable depth using un-structured meshes*, Comput. Methods Appl. Mech. Engrg. **155** (1998), 49-72.

[14] R. Botchorishvili, B. Perthame, A. Vasseur, *Eqilibrium schemes for scalar conservation laws with stiff sources*, Math. Comp. **72** (2003), 131-157.

[15] N. Botta, R. Klein, S. Langenberg, S. Lützenkirchen, *Well-balanced finite volume methods for nearly hydrostatic flows*, submitted, March 2002.

[16] F. Bouchut, *Construction of BGK models with a family of kinetic entropies for a given system of conservation laws*, J. Stat. Phys. **95** (1999), 113-170.

[17] F. Bouchut, *Entropy satisfying flux vector splittings and kinetic BGK models*, Numer. Math. **94** (2003), 623-672.

[18] F. Bouchut, *A reduced stability condition for nonlinear relaxation to conservation laws*, J. Hyp. Diff. Eq. **1** (2004), 149-170.

[19] F. Bouchut, J. LeSommer, V. Zeitlin, *Frontal geostrophic adjustment and nonlinear-wave phenomena in one dimensional rotating shallow water; Part 2 : high resolution numerical investigation*, to appear in J. Fluid Mech. (2004).

[20] F. Bouchut, A. Mangeney-Castelnau, B. Perthame, J.-P. Vilotte, *A new model of Saint Venant and Savage-Hutter type for gravity driven shallow water flows*, C. R. Acad. Sci. Paris Sér. I Math. **336** (2003), 531-536.

[21] F. Bouchut, T. Morales, *A subsonic-well-balanced reconstruction scheme for shallow water flows*, preprint 2004.

[22] F. Bouchut, M. Westdickenberg, *Gravity driven shallow water models for arbitrary topography*, preprint 2004.

[23] Y. Brenier, *Averaged multivalued solutions for scalar conservation laws*, SIAM J. Numer. Anal. **21** (1984), 1013-1037.

[24] T. Buffard, T. Gallouët, J. M. Hérard, *A sequel to a rough Godunov scheme: application to real gases*, Computers and Fluids **29** (2000), 813-847.

[25] M. Castro, J. Macías, C. Parés, *A Q-scheme for a class of systems of coupled conservation laws with source term, application to a two-layer 1-D shallow water system*, M2AN Math. Model. Numer. Anal. **35** (2001), 107-127.

[26] C. Chalons, Thesis Ecole Polytechnique Palaiseau, France, 2002.

[27] G.Q. Chen, C.D. Levermore, T.-P. Liu, *Hyperbolic conservation laws with stiff relaxation terms and entropy*, Comm. Pure Appl. Math. **47** (1994), 787-830.

[28] A. Chinnayya, A.-Y. Le Roux, *A new general Riemann solver for the shallow-water equations with friction and topography*, preprint.

[29] A. Chinnayya, A.-Y. LeRoux, N. Seguin, *A well-balanced numerical scheme for the approximation of the shallow-water equations with topography: the resonance phenomenon*, to appear in Int. J. Finite Volume, 2004.

[30] F. Coquel, E. Godlewski, B. Perthame, A. In, P. Rascle, *Some new Godunov and relaxation methods for two-phase flow problems*, Godunov methods (Oxford, 1999), 179-188, Kluwer/Plenum, New York, 2001.

[31] C.M. Dafermos, *Hyperbolic conservation laws in continuum physics*, Grundlehren der Mathematischen Wissenschaften [Fundamental Principles of Mathematical Sciences], 325, Springer-Verlag, Berlin, 2000.

[32] B. Després, *Lagrangian systems of conservation laws. Invariance properties of Lagrangian systems of conservation laws, approximate Riemann solvers and the entropy condition*, Numer. Math. **89** (2001), 99-134.

[33] R. Eymard, T. Gallouët, R. Herbin, *Finite volume methods*, In Handbook of Numerical Analysis (Vol. VII), P.G. Ciarlet and J.-L. Lions editors, North-Holland, 2000.

[34] F. Filbet, C.-W. Shu, *Approximation of hyperbolic models for chemosensitive movement*, preprint 2004.

[35] H. Frid, *Maps of convex sets and invariant regions for finite-difference systems of conservation laws*, Arch. Ration. Mech. Anal. **160** (2001), 245-269.

[36] H. Frid, *Correction to "Maps of convex sets and invariant regions for finite-difference systems of conservation laws"*, Arch. Ration. Mech. Anal. **171** (2004), 297-299.

[37] G. Gallice, *Positive and entropy stable Godunov-type schemes for gas dynamics and MHD equations in Lagrangian or Eulerian coordinates*, Numer. Math. **94** (2003), 673-713.

[38] T. Gallouët, J.-M. Hérard, N. Seguin, *Some approximate Godunov schemes to compute shallow-water equations with topography*, Computers and Fluids **32** (2003), 479-513.

[39] T. Gallouët, J.-M. Hérard, N. Seguin, *Some recent finite volume schemes to compute Euler equations using real gas EOS*, Internat. J. Numer. Methods Fluids **39** (2002), 1073-1138.

[40] P. García-Navarro, M.E. Vázquez-Cendón, *On numerical treatment of the source terms in the shallow water equations*, Comput. Fluids **29** (2000), 17-45.

[41] T. Gimse, N.H. Risebro, *Riemann problems with a discontinuous flux function*, In Proc. 3rd Internat. Conf. Hyperbolic Problems, 488-502, Uppsala, 1991, Studentlitteratur.

[42] P. Goatin, P.G. LeFloch, *The Riemann problem for a class of resonant hyperbolic systems of balance Laws*, preprint 2003.

[43] G. Godinaud, A.-Y. LeRoux, M.-N. LeRoux, *Generation of new solvers involving the source term for a class of nonhomogeneous hyperbolic systems*, preprint.

[44] E. Godlewski, P.-A. Raviart, *Hyperbolic systems of conservation laws*, Mathématiques & Applications 3/4, Ellipses, Paris, 1991.

[45] E. Godlewski, P.-A. Raviart, *Numerical approximation of hyperbolic systems of conservation laws*, Applied Mathematical Sciences, 118, Springer-Verlag, New York, 1996.

[46] L. Gosse, *A well-balanced flux-vector splitting scheme designed for hyperbolic systems of conservation laws with source terms*, Comp. Math. Appl. **39** (2000), 135-159.

[47] L. Gosse, *Localization effects and measure source terms in numerical schemes for balance laws*, Math. Comp. **71** (2002), 553-582.

[48] L. Gosse, *A well-balanced scheme using non-conservative products designed for hyperbolic systems of conservation laws with source terms*, Math. Mod. Meth. Appl. Sci. **11** (2001), 339-365.

[49] L. Gosse, A.-Y. LeRoux, *A well-balanced scheme designed for inhomogeneous scalar conservation laws*, C. R. Acad. Sci. Paris Sér. I Math. **323** (1996), 543-546.

[50] L. Gosse, G. Toscani, *Space localization and well-balanced schemes for discrete kinetic models in diffusive regimes*, SIAM J. Numer. Anal. **41** (2003), 641-658.

[51] N. Goutal, F. Maurel, *Proceedings of the second workshop on dam-break wave simulation*, Technical report EDF/DER HE-43/97/016/B, Châtou, France (1997).

[52] J.M. Greenberg, A.-Y. LeRoux, *A well-balanced scheme for the numerical processing of source terms in hyperbolic equations*, SIAM J. Numer. Anal. **33** (1996), 1-16.

[53] J.-M. Greenberg, A.-Y. Leroux, R. Baraille, A. Noussair, *Analysis and approximation of conservation laws with source terms*, SIAM J. Numer. Anal. **34** (1997), 1980-2007.

[54] G. Guerra, *Well-posedness fo a scalar conservation law with singular nonconservative source*, preprint 2004.

[55] P. Gwiazda, *An existence result for a model of granular material with nonconstant density*, Asympt. Anal. **30** (2002), 43-60.

[56] A. Harten, P.D. Lax, B. Van Leer, *On upstream differencing and Godunov-type schemes for hyperbolic conservation laws*, SIAM Review **25** (1983), 35-61.

[57] E. Isaacson, B. Temple, *The structure of asymptotic states in a singular system of conservation laws*, Adv. Appl. Math. **11** (1990), 205-219.

[58] E. Isaacson, B. Temple, *Nonlinear resonance in systems of conservation laws*, SIAM J. Appl. Math. **52** (1992), 1260-1278.

[59] E. Isaacson, B. Temple, *Convergence of the 2×2 Godunov method for a general resonant nonlinear balance law*, SIAM J. Appl. Math. **55** (1995), 625-640.

[60] S. Jin, *Efficient asymptotic-preserving (AP) schemes for some multiscale kinetic equations*, SIAM J. Sci. Comput. **21** (1999), 441-454.

[61] S. Jin, *A steady-state capturing method for hyperbolic systems with geometrical source terms*, Math. Model. Numer. Anal. **35** (2001), 631-645.

[62] S. Jin, L. Pareschi, G. Toscani, *Uniformly accurate diffusive relaxation schemes for multiscale transport equations*, SIAM J. Numer. Anal. **38** (2000), 913-936.

[63] S. Jin, Z.-P. Xin, *The relaxation schemes for systems of conservation laws in arbitrary space dimensions*, Comm. Pure Appl. Math. **48** (1995), 235-276.

[64] K.H. Karlsen, N.H. Risebro, J.D. Towers, *L 1 stability for entropy solutions of nonlinear degenerate parabolic convection-diffusion equations with discontinuous coefficients*, Skr. K. Nor. Vidensk. Selsk. **3** (2003), 49.

[65] T. Katsaounis, B. Perthame, C. Simeoni, *Upwinding Sources at Interfaces in conservation laws*, to appear in Appl. Math. Lett.

[66] T. Katsaounis, C. Simeoni, *First and second order error estimates for the upwind interface source method*, preprint 2002.

[67] T. Katsaounis, C. Simeoni, *Second order approximation of the viscous Saint Venant system and comparison with experiments*, Proceedings of Hyp2002, T. Hou and E. Tadmor editors, Springer, 2003.

[68] A. Klar, A. Unterreiter, *Uniform stability of a finite difference scheme for transport equations in diffusive regimes*, SIAM J. Numer. Anal. **40** (2002), 891-913.

[69] C. Klingenberg, N. H. Risebro, *Conservation laws with discontinuous coefficients, existence, uniqueness and asymptotic behavior*, Comm. Partial Differential Equations **20** (1995), 1959-1990.

[70] C. Klingenberg, N. H. Risebro, *Stability of a resonant system of conservation laws modeling polymer flow with gravitation*, J. Differential Equations **170** (2001), 344-380.

[71] B. Larrouturou, *How to preserve the mass fractions positivity when comput- ing compressible multi-component flows*, J. Comput. Phys. **95** (1991), 59-84.

[72] P.G. LeFloch, *Hyperbolic systems of conservation laws, the theory of classical and nonclassical shock waves*, ETH Lecture Notes Series, Birkhauser, 2002.

[73] P.G. LeFloch, M.D. Thanh, *The Riemann problem for fluid flows in a nozzle with discontinuous cross section*, Comm. Math. Sci. **1** (2003), 763-797.

[74] A.-Y. LeRoux, *Riemann solvers for some hyperbolic problems with a source term*, ESAIM proc. **6** (1999), 75-90.

[75] J. LeSommer, G. Reznik, V. Zeitlin, *Nonlinear geostrophic adjustment of long-wave disturbances in the shallow water model on the equatorial beta- plane*, preprint 2003.

[76] R.J. LeVeque, *Balancing source terms and flux gradients in high-resolution Godunov methods: the quasi-steady wave-propagation algorithm*, J. Comp. Phys. **146** (1998), 346-365.

[77] R.J. LeVeque, *Finite volume methods for hyperbolic problems*, Cambridge Univ. Press, 2002.

[78] R.J. LeVeque, M. Pelanti, *A class of approximate Riemann solvers and their relation to relaxation schemes*, J. Comput. Phys. **172** (2001), 572-591.

[79] L.W. Lin, B.J. Temple, J.H. Wang, *A comparison of convergence rates for Godunov's method and Glimm's method in resonant nonlinear systems of conservation laws*, SIAM J. Numer. Anal. **32** (1995), 824-840.

[80] L.W. Lin, B.J. Temple, J.H. Wang, *Suppression of oscillations in Godunov's method for a resonant non-strictly hyperbolic system*, SIAM J. Numer. Anal. **32** (1995), 841-864.

[81] T.P. Liu, *Quasilinear hyperbolic systems*, Comm. Math. Phys. **68** (1979), 141-172.

[82] T.-P. Liu, *Nonlinear resonance for quasilinear hyperbolic equation*, J. Math. Phys. **28** (1987), 2593-2602.

[83] A. Mangeney-Castelnau, F. Bouchut, E. Lajeunesse, A. Aubertin, M. Pir- ulli, J.-P. Vilotte, *On the use of Saint Venant equations for simulating the spreading of a granular mass*, preprint 2004.

[84] C. Mascia, A. Terracina, *Large-time behavior for conservation laws with source in a bounded domain*, J. Diff. Eq. **159** (1999), 485-514.

[85] G. Naldi, L. Pareschi, *Numerical schemes for hyperbolic systems of conser- vation laws with stiff diffusive relaxation*, SIAM J. Numer. Anal. **37** (2000), 1246-1270.

[86] B. Perthame, *Kinetic formulations of conservation laws*, Oxford University Press, 2002.

[87] B. Perthame, C. Simeoni, *A kinetic scheme for the Saint Venant system with a source term*, Calcolo **38** (2001), 201-231.

[88] B. Perthame, C. Simeoni, *Convergence of the upwind interface source method for hyperbolic conservation laws*, Proc. of Hyp2002, T. Hou and E. Tadmor Editors, Springer, 2003.

[89] P.L. Roe, *Approximate Riemann solvers, parameter vectors and difference schemes*, J. Comp. Phys. **43** (1981), 357-372.

[90] N. Seguin, J. Vovelle, *Analysis and approximation of a scalar conservation law with a flux function with discontinuous coefficients*, Math. Models and Meth. in Appl. Sci. (M3AS) **13** (2003), 221-250.

[91] D. Serre, *Systems of conservation laws. 1. Hyperbolicity, entropies, shock waves*, Translated from the 1996 French original by I. N. Sneddon, Cambridge University Press, Cambridge, 1999.

[92] D. Serre, *Systems of conservation laws. 2. Geometric structures, oscillations, and initial-boundary value problems*, Translated from the 1996 French original by I. N. Sneddon, Cambridge University Press, Cambridge, 2000.

[93] I. Suliciu, *On modelling phase transitions by means of rate-type constitutive equations, shock wave structure*, Internat. J. Engrg. Sci. **28** (1990), 829-841.

[94] I. Suliciu, *Some stability-instability problems in phase transitions modelled by piecewise linear elastic or viscoelastic constitutive equations*, Internat. J. Engrg. Sci. **30** (1992), 483-494.

[95] E. Tadmor, *Entropy stability theory for difference approximations of nonlinear conservation laws and related time dependent problems*, Acta Numerica (2003), 451-512.

[96] B. Temple, *Global solution of the Cauchy problem for a class of* 2×2 *non-strictly hyperbolic conservation laws*, Adv. Appl. Math. **3** (1982), 335-375.

[97] E.F. Toro, *Riemann solvers and numerical methods for fluid dynamics: a practical introduction*, second edition, Springer-Verlag, 1999.

[98] J.D. Towers, *Convergence of a difference scheme for conservation laws with a discontinuous flux*, SIAM J. Numer. Anal. **38** (2000), 681-698.

[99] A. Vasseur, *Well-posedness of scalar conservation laws with singular sources*, Methods Appl. Anal. **9** (2002), 291-312.

[100] M.E. Vázquez-Cendón, *Improved treatment of source terms in upwind schemes for the shallow water equations in channels with irregular geometry*, J. Comput. Phys. **148** (1999), 497-526.

[101] J. Wang, H. Wen, T. Zhou, *On large time step Godunov scheme for hyperbolic conservation laws*, preprint 2004.

Index

Frontiers in Mathematics

*Your Specialized
Publisher in
Mathematics*

Birkhäuser

Further titles

■ **Kasch, F.**, Universität München, Germany / **Mader, A.**,
Hawaii University
Rings, Modules, and the Total
2004. 148 pages. Softcover
ISBN 3-7643-7125-0

In a nutshell, this monograph deals with direct
decompositions of modules and associated concepts. The
central notion of "partially invertible homomorphisms",
namely those that are factors of a non-zero idempotent, is
introduced in a very accessible fashion. Units and regular
elements are partially invertible. The "total" consists of all
elements that are not partially invertible. The total contains
the radical and the singular and cosingular submodules,
but while the total is closed under right and left
multiplication, it may not be closed under addition. Cases
are discussed where the total is additively closed. The total
is particularly suited to deal with the endomorphism ring of
a direct sum of modules that all have local endomorphism
rings and is applied in this case. Further applications are
given for torsion-free Abelian groups.
This book offers for the first time a comprehensive and
readable exposition of results on the total. Although
dealing with recent research, the material is accessible to
anyone with a basic knowledge of ring and module theory.
A short introduction to torsion-free Abelian groups is
included. The subject is by no means exhausted and topics
for further research can easily be found.

■ **Thas, K.**, Ghent University, Ghent, Belgium
Symmetry in Finite Generalized Quadrangles
2004. 240 pages. Softcover
ISBN 3-7643-6158-1

In this book, a classification of finite generalized
quadrangles based on the possible subconfigurations of
axes of symmetry is proposed, extending thus the
celebrated Lenz-Barlotti classification for projective planes
to the theory of generalized quadrangles.
Several open problems and long-standing conjectures are
solved, respectively answered, by new techniques arising
from a mixture of geometrical, combinatorial and group
theoretical arguments. Many new, previously unpublished
results with proofs are presented.

The book is aimed at advanced graduate students and
researchers in the area. Readers will find a self-contained
introduction to the modern theory of finite generalized
quadrangles and related structures, as well as a detailed
account of the classification and its implications.

■ **Krausshar, R.S.**, Ghent University, Ghent, Belgium
**Generalized Analytic Automorphic Forms in
Hypercomplex Spaces**
2004. 182 pages. Softcover
ISBN 3-7643-7059-9

The aim of this book is to provide a first comprehensive
overview of the basic theory of hypercomplex-analytic
automorphic forms and functions for arithmetic subgroups
of the Vahlen group in higher dimensional spaces. It gives a
summary on the research results obtained over the last five
years and establishes a new field within the theory of
functions of hypercomplex variables and within analytic
number theory.
Hypercomplex-analyticity generalizes the concept of
complex analyticity in the sense of considering
null-solutions to higher dimensional Cauchy-Riemann type
systems. Vector- and Clifford algebra-valued Eisenstein and
Poincaré series are constructed within this framework and a
detailed description of their analytic and number
theoretical properties is provided. In particular, explicit
relationships to higher dimensional vector valued variants
of the Riemann zeta function and Dirichlet series are
established and a concept of hypercomplex multiplication
of lattices is introduced. Applications to the theory of
Hilbert spaces with reproducing kernels, to partial
differential equations and index theory on some
conformally flat manifolds are also included.
The book is directed to researchers as well as to graduate
and postgraduate students with interest in the fields of the
theory of generalized analytic functions in higher
dimensional spaces, analytic number theory, function
spaces and boundary value problems of partial differential
equations on conformally flat manifolds, and some closely
related fields in physics, such as instanton theory and
quantum gravity.